INTRODUCTION TO STATISTICS THROUGH RESAMPLING METHODS AND R/S-PLUS®

INTRODUCTION TO STATISTICS THROUGH RESAMPLING METHODS AND R/S-PLUS®

Phillip I. Good

Huntington Beach, California

A JOHN WILEY & SONS, INC., PUBLICATION

Library of Congress Cataloging-in-Publication Data:

Good, Phillip I.
    Introduction to statistics through resampling methods and R/S-PLUS / Phillip I. Good.
        p. cm.
    Includes bibliographical references and indexes.
    ISBN-13 978-0-471-71575-7 (pbk. : acid-free paper)
    ISBN-10 0-471-71575-1 (pbk. : acid-free paper)
    1. Resampling (Statistics) 2. S-Plus. I. Title.

    QA278.8.G63 2005
    519.5′4—dc22   2004061922

Printed in the United States of America

10 9 8 7 6 5 4 3 2

CONTENTS

Preface   xi

1 Variation   1

1.1 Variation / 1
1.2 Collecting Data / 3
1.3 Summarizing Your Data / 4
    1.3.1 Learning to Use R / 4
1.4 Reporting Your Results / 7
    1.4.1 Picturing Data / 8
    1.4.2 Better Graphics / 10
1.5 Types of Data / 11
    1.5.1 Depicting Categorical Data / 12
1.6 Displaying Multiple Variables / 12
    1.6.1 From Observations to Questions / 14
1.7 Measures of Location / 14
    1.7.1 Which Measure of Location? / 15
    1.7.2 The Bootstrap / 19
1.8 Samples and Populations / 21
    1.8.1 Drawing a Random Sample / 22
    1.8.2 Using R to Draw a Sample / 23
    1.8.3 Ensuring the Sample Is Representative / 25
1.9 Variation—Within and Between / 25
1.10 Summary and Review / 27

2 Probability   29

2.1 Probability / 29
    2.1.1 Events and Outcomes / 31

v

2.1.2   Venn Diagrams / 31

2.2   Binomial / 33

    2.2.1   Permutations and Rearrangements / 35
    2.2.2   Back to the Binomial / 38
    2.2.3   The Problem Jury / 38
    2.2.4   Properties of the Binomial / 39
    2.2.5   Multinomial / 43

2.3   Conditional Probability / 43

    2.3.1   Market Basket Analysis / 45
    2.3.2   Negative Results / 46

2.4   Independence / 47
2.5   Applications to Genetics / 49
2.6   Summary and Review / 50

3 Distributions   **52**

3.1   Distribution of Values / 52

    3.1.1   Cumulative Distribution Function / 53
    3.1.2   Empirical Distribution Function / 54

3.2   Discrete Distributions / 55
3.3   Poisson: Events Rare in Time and Space / 58

    3.3.1   Applying the Poisson / 58
    3.3.2   Comparing Empirical and Theoretical Poisson Distributions / 59

3.4   Continuous Distributions / 60

    3.4.1   The Exponential Distribution / 61
    3.4.2   Normal Distribution / 62
    3.4.3   Mixtures of Normal Distributions / 64

3.5   Properties of Independent Observations / 64
3.6   Testing a Hypothesis / 66

    3.6.1   Analyzing the Experiment / 67
    3.6.2   Two Types of Errors / 69

3.7   Estimating Effect Size / 71

    3.7.1   Additional Applications / 71
    3.7.2   Using Confidence Intervals to Test Hypotheses / 73

3.8   Summary and Review / 74

4 Testing Hypotheses   **76**

4.1   One-Sample Problems / 76

   4.1.1   Percentile Bootstrap / 76
   4.1.2   Parametric Bootstrap / 77
   4.1.3   Student's *t* / 78

4.2   Comparing Two Samples / 80

   4.2.1   Comparing Two Poisson Distributions / 80
   4.2.2   What Should We Measure? / 80
   4.2.3   Permutation Monte Carlo / 81
   4.2.4   One- versus Two-Sided Tests / 83
   4.2.5   Bias-Corrected Nonparametric Bootstrap / 83
   4.2.6   Two-Sample *t*-Test / 86

4.3   Which Test Should We Us? / 87

   4.3.1   *p*-Values and Significance Levels / 87
   4.3.2   Test Assumptions / 88
   4.3.3   Robustness / 89
   4.3.4   Power of a Test Procedure / 90
   4.3.5   Testing for Correlation / 92

4.4   Summary and Review / 94

5 Designing an Experiment or Survey   **96**

5.1   The Hawthorne Effect / 97

   5.1.1   Crafting an Experiment / 98

5.2   Designing an Experiment or Survey / 100

   5.2.1   Objectives / 100
   5.2.2   Sample from the Right Population / 101
   5.2.3   Coping with Variation / 103
   5.2.4   Matched Pairs / 104
   5.2.5   The Experimental Unit / 105
   5.2.6   Formulate Your Hypotheses / 106
   5.2.7   What Are You Going to Measure / 107
   5.2.8   Random Representative Samples / 108
   5.2.9   Treatment Allocation / 109
   5.2.10   Choosing a Random Sample / 110
   5.2.11   Ensuring Your Observations Are
            Independent / 111

5.3   How Large a Sample? / 112

   5.3.1   Samples of Fixed Size / 113

Known Distribution / 114
Almost Normal Data / 117
Bootstrap / 119
5.3.2 Sequential Sampling / 120
Stein's Two-Stage Sampling Procedure / 120
Wald Sequential Sampling / 121
Adaptive Sampling / 126
5.4 Meta-analysis / 126
5.5 Summary and Review / 126

6 Analyzing Complex Experiments   **129**

6.1 Changes Measured in Percentages / 129
6.2 Comparing More Than Two Samples / 130
6.2.1 Programming the Multisample Comparison
in R / 131
6.2.2 Reusing Your R Functions / 133
6.2.3 What Is the Alternative? / 133
6.2.4 Testing for a Dose Response or Other Ordered
Alternative / 134
6.3 Equalizing Variances / 137
6.4 Categorical Data / 139
6.4.1 One-Sided Fisher's Exact Test / 142
6.4.2 The Two-Sided Test / 143
6.4.3 Multinomial Tables / 144
6.5 Multivariate Analysis / 145
6.5.1 Manipulating Multivariate Data in R / 146
6.5.2 Pesarin–Fisher Omnibus Statistic / 147
6.5.3 Programming Guidelines / 149
6.6 Summary and Review / 154

7 Developing Models   **155**

7.1 Models / 155
7.1.1 Why Build Models? / 156
7.1.2 Caveats / 158
7.2 Regression / 158
7.2.1 Linear Regression / 160
7.3 Fitting a Regression Equation / 161
7.3.1 Ordinary Least Squares / 161
Types of Data / 165

7.3.2   Least Absolute Deviation Regression / 165
7.3.3   Errors-in-Variables Regression / 166
7.3.4   Assumptions / 168
7.4   Problems with Regression / 169
7.4.1   Goodness of Fit Versus Prediction / 170
7.4.2   Which Model? / 170
Measures of Predictive Success / 172
7.4.3   Multivariable Regression / 172
7.5   Quantile Regression / 181
7.6   Validation / 183
7.6.1   Independent Verification / 184
7.6.2   Splitting the Sample / 184
7.6.3   Cross-validation with the Bootstrap / 186
7.7   Classification and Regression Trees (CART) / 186
7.8   Summary and Review / 190

8   Reporting Your Findings   192

8.1   What to Report / 193
8.2   Text, Table, or Graph? / 196
8.3   Summarizing Your Results / 197
8.3.1   Center of the Distribution / 199
8.3.2   Dispersion / 201
8.4   Reporting Analysis Results / 202
8.4.1   *p*-Values or Confidence Intervals? / 203
8.5   Exceptions Are the Real Story / 205
8.5.1   Nonresponders / 205
8.5.2   The Missing Holes / 205
8.5.3   Missing Data / 206
8.5.4   Recognize and Report Biases / 206
8.6   Summary and Review / 207

9   Problem Solving   208

9.1   The Problems / 208
9.2   Solving Practical Problems / 213
9.2.1   The Data's Provenance / 213
9.2.2   Inspect the Data / 213
9.2.3   Validate the Data Collection Methods / 214
9.2.4   Formulate Hypotheses / 215

9.2.5 Choosing a Statistical Methodology / 215

9.2.6 Be Aware of What You Don't Know / 216

9.2.7 Qualify Your Conclusions / 216

Appendix: S-PLUS   **218**

Index to R Commands   **225**

Index   **227**

PREFACE

Intended for class use or self-study, this text aspires to introduce statistical methodology to a wide audience, simply and intuitively, through resampling from the data at hand.

The resampling methods—permutations and the bootstrap—are easy to learn and easy to apply. They require no mathematics beyond introductory high-school algebra, yet are applicable in an exceptionally broad range of subject areas.

Introduced in the 1930s, the numerous, albeit straightforward calculations that resampling methods require were beyond the capabilities of the primitive calculators then in use. They were soon displaced by less powerful, less accurate approximations that made use of tables. Today, with a powerful computer on every desktop, resampling methods have resumed their dominant role and table lookup is an anachronism.

Physicians and physicians in training, nurses and nursing students, business persons, business majors, and research workers and students in the biological and social sciences will find here a practical and easily grasped guide to descriptive statistics, estimation, testing hypotheses, and model building.

For advanced students in biology, dentistry, medicine, psychology, sociology, and public health, this text can provide a first course in statistics and quantitative reasoning.

For mathematics majors, this text will form the first course in statistics to be followed by a second course devoted to distribution theory and asymptotic results.

Hopefully, all readers will find my objectives are the same as theirs: *to use quantitative methods to characterize, review, report on, test, estimate, and classify findings*.

Warning to the autodidact: You can master the material in this text without the aid of an instructor. But you may not be able to grasp even the more elementary concepts without completing the exercises. Whenever and wherever you encounter an exercise in the text, stop your reading and complete the exercise before going further. To simplify the task, R code and data sets may be downloaded from `ftp://ftp.wiley.com/public/sci_tech_med/statistics_resampling/` and then cut and pasted into your programs.

The S programming language is used to illustrate the concepts in this text and to aid readers in completing the exercises. S is available in two flavors, R and S-PLUS®. R may be downloaded without charge for use under Windows, UNIX, or the Macintosh from `http://cran.r-project.org`. S-PLUS is available for Windows or UNIX from `http://www.insightful.com`, with a free student version from `http://elms03.e-academy.com/splus`. R and S-PLUS share a programming language (with minor differences); in addition, S-PLUS has a graphical user interface, and a library that simplifies resampling. An introduction to these features is provided in an appendix.

For a one-quarter short course, I took the students through Chapters 1, 2, and part of 3. We completed Chapters 3 and 4 in the winter quarter and started Chapter 5, finishing the year with Chapters 5, 6, and 7. Chapters 8 and 9, Reports your Problems and Problem Solving, convert the text into an invaluable professional resource.

Twenty-eight or more exercises included in each chapter plus dozens of thought-provoking questions in Chapter 9 will serve the needs of both classroom and self-study. The discovery method is utilized as often as possible and the student and conscientious reader are forced to think their way to a solution rather than being able to copy the answer or apply a formula straight out of the text.

If you find this text an easy read, then your gratitude should go to Cliff Lunneborg for his many corrections and clarifications. Tim Hesterberg made many essential suggestions. I am deeply indebted to both Cliff Lunneborg and Rob J. Goedman for their help with the R language, and to Michael L. Richardson, David Warton, Mike Moreau, Lynn Marek, Mikko Mönkkönen, Kim Colyvas, my students at UCLA, and the students in the Introductory Statistics and Resampling Methods courses that I offer online each quarter through the auspices of `statistics.com` for their comments and corrections.

An Instructor's Manual may be obtained by contacting the Publisher. Please visit `ftp://ftp.wiley.com/public/sci_tech_med/statistics_resampling/` for instructions on how to request a copy of the manual.

PHILLIP I. GOOD

Huntington Beach, California
`frere_until@hotmail.com`

1

VARIATION

If there were no variation, if every observation were predictable, a mere repetition of what had gone before, there would be no need for statistics.

In this chapter, you'll learn what statistics is all about and how to use R to display data you've collected.

1.1. VARIATION

I find physics extremely satisfying. In high school, we learned the formula $S = VT$, which in symbols relates the distance traveled by an object to its velocity multiplied by the time spent in traveling. If the speedometer indicates 60 miles an hour, then in half an hour you are certain to travel exactly 30 miles. Except that during our morning commute, the speed we travel is seldom constant.

In college, we had Boyle's Law, $V = KT/P$, with its tidy relationship between the volume V, temperature T, and pressure P of a perfect gas. This is just one example of the perfection encountered there. The problem was we could never quite duplicate this (or any other) law in the freshman physics laboratory. Maybe it was the measuring instruments, our lack of

Introduction to Statistics Through Resampling Methods and R/S-PLUS®, By Phillip I. Good
Copyright © 2005 by John Wiley & Sons, Inc.

familiarity with the equipment, or simple measurement error, but we kept getting different values for the constant K.

By now, we know that variation is the norm. Instead of getting a fixed, reproducible V to correspond to a specific T and P, one ends up with a distribution of values instead as a result of errors in measurement. But we also know that with a large enough sample, the mean and shape of this distribution are reproducible.

That's the good news: Make astronomical, physical, or chemical measurements and the only variation appears to be due to observational error. But try working with people.

Anyone who has spent any time in a schoolroom, whether as a parent or as a child, has become aware of the vast differences among individuals. Our most distinct memories are of how large the girls were in the third grade (ever been beat up by a girl?) and the trepidation we felt on the playground whenever teams were chosen (not right field again!). Much later, in our college days, we were to discover there were many individuals capable of devouring larger quantities of alcohol than we could without noticeable effect. And a few, mostly of other nationalities, whom we could drink under the table.

Whether or not you imbibe, we're sure you've had the opportunity to observe the effects of alcohol on others. Some individuals take a single drink and their noses turn red. Others can't seem to take just one drink.

Despite these obvious differences, scheduling for out-patient radiology at many hospitals is done by a computer program that allots exactly 15 minutes to each patient. Well, I've news for them and their computer. Occasionally, the technologists are left twidling their thumbs. More often the waiting room is overcrowded because of routine exams that weren't routine or where the radiologist wanted additional shots. (To say nothing of those patients who show up an hour or so early or a half hour late.)

The majority of effort in *experimental design*, the focus of Chapter 5 of this text, is devoted to finding ways in which this variation from individual to individual won't swamp or mask the variation that results from differences in treatment or approach. It's probably safe to say that what distinguishes statistics from all other branches of applied mathematics is that it is devoted to characterizing and then accounting for *variation*.

CONSIDER THE FOLLOWING EXPERIMENT

You catch three fish. You heft each one and estimate its weight; you weigh each one on a pan scale when you get back to dock, and you take them to a chemistry laboratory and weigh them there. Your two friends on the boat do exactly the same thing. (All but Mike; the chem professor catches him and calls campus security. This is known as missing data.)

The 26 weights you've recorded ($3 \times 3 \times 3 - 1$ when they nabbed Mike) differ as the result of measurement error, observer error, differences among observers, differences among measuring devices, and differences among fish.

1.2. COLLECTING DATA

The best way to observe variation is for you, the reader, to collect some data. But before we make some suggestions, a few words of caution are in order: 80% of the effort in any study goes into data collection and preparation for data collection. Any effort you don't expend goes into cleaning up the resulting mess.

We constantly receive letters and e-mails asking which statistic we would use to rescue a misdirected study. There is no magic formula, no secret procedure known only to Ph.D. statisticians. The operative phrase is GIGO: garbage in, garbage out. So think carefully before you embark on your collection effort. Make a list of possible sources of variation and see if you can eliminate any that are unrelated to the objectives of your study. If midway through, you think of a better method—don't use it. Any inconsistency in your procedure will only add to the undesired variation.

Let's get started. Here are three suggestions. Before continuing with your reading, follow through on at least one of them or an equivalent idea of your own as we will be using the data you collect in the very next section.

1. Measure the height, circumference, and weight of a dozen humans (or dogs, or hamsters, or frogs, or crickets).
2. Time some tasks. Record the times of five to ten individuals over three track lengths (say, 50 meters, 100 meters, and a quarter mile). Since the participants (or trial subjects) are sure to complain they could have done much better if only given the opportunity, record at least two times for each study subject. (Feel free to use frogs, hamsters, or

turtles in place of humans as runners to be timed; or to replace footraces with knot tying, bandaging, or putting on a uniform.)

3. Take a survey. Include at least three questions and survey at least ten subjects. All your questions should take the form: "Do you prefer A to B? Strongly prefer A, slightly prefer A, indifferent, slightly prefer B, strongly prefer B." For example, "Do you prefer Britney Spears to Jennifer Lopez?" or "Would you prefer spending money on new classrooms rather than guns?"

Exercise 1.1. Collect data as described above. Before you begin, write down a complete description of exactly what you intend to measure and how you plan to make your measurements. Make a list of all potential sources of variation. When your study is complete, describe what deviations you had to make from your plan and what additional sources of variation you encountered.

1.3. SUMMARIZING YOUR DATA

Learning how to adequately summarize one's data can be a major challenge. Can it be explained with a single number like the median or mean? The *median* is the middle value of the observations you have taken, so that half the data has a smaller value and half has a greater value. Take the observations 1.2, 2.3, 4.0, 3, and 5.1. The observation 3 is the one in the middle. If we have an even number of observations such as 1.2, 2.3, 3, 3.8, 4.0, and 5.1, then the best one can say is that the median or midpoint is a number (any number) between 3 and 3.8. Now, a question for you: What are the median values of the measurements you made in our first exercise?

Hopefully, you've already collected data as described in the preceding section; otherwise, face it, you are behind. Get out the tape measure and the scales. If you conducted time trials, use those data instead. Treat the observations for each of the three distances separately.

If you conducted a survey, we have a bit of a problem. How does one translate "I would prefer spending money on new classrooms rather than guns" into a number a computer can add and subtract? There is more than one way to do this, as we'll discuss in Section 1.5, Types of Data. For the moment, assign the number 1 to "strongly prefer classrooms," the number 2 to "slightly prefer classrooms," and so on.

1.3.1. Learning to Use R

Calculating the value of a statistic is easy enough when we've only one or two observations, but a major pain when we have ten or more. And as for drawing graphs—one of the best ways to summarize your data—we're no artists. Let the computer do the work.

We're going to need the help of a programming language R that is specially designed for use in computing statistics and creating graphs. You can download that language without charge from the website `http://cran.r-project.org/`.

R is an interpreter. This means that as we enter the lines of a typical program, we'll learn on a line-by-line basis whether the command we've entered makes sense (to the computer) and be able to correct the line if we've made a typing error.

When we run R, what we see on the screen is an arrowhead ➣.

If we type 2 + 3 after and then press the enter key, we see [1] 5. (R reports numeric results in the form of a vector. In this example, the first and only element in this vector takes the value 5.)

To enter the observations 1.2, 2.3, 4.0, 3, and 5.1, type

```
ourdata = c(1.2, 2.3, 4.0, 3, 5.1)
```

If you've never used a programming language before, let me warn you that R is very inflexible. It won't understand (or, worse, may misinterpret) all of the following:

```
ourdata = c(1.2 2.3 4.0 3 5.1)
ourdata = (1.2, 2.3, 4.0, 3, 5.1)
```

If you did type the line correctly, then typing median(ourdata) after will yield the answer 3 after you hit the enter key.

```
ourdata = c(1.2 2.3 4.0 3 5.1)
Error: syntax error
ourdata = c(1.2, 2.3, 4.0, 3, 5.1)
median(ourdata)
[1] 3
➣
```

The median may tell us where the center of a distribution is, but it provides no information about the variability of our observations, and variation is what statistics is all about. Pictures tell the story the best.

The one-way *strip chart* (Figure 1.1)[1] reveals that the *minimum* of this particular set of data is 0.9 and the *maximum* is 24.8. Each vertical line in this strip chart corresponds to an observation. Darker lines correspond to multiple observations. The *range* over which these observations extend is 24.8 − 0.9 or about 24.

---

[1] R code for Figures 1.1 to 1.3 is provided in Section 1.4.1.

Figure 1.1. Strip chart.

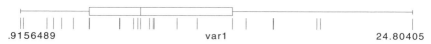

Figure 1.2. Combination *box plot* (top section) and one-way strip chart.

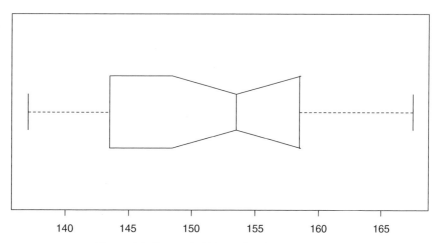

Figure 1.3. Box and whiskers plot of classroom data.

Figure 1.2 shows a combination *box plot* (top section) and one-way strip chart (lower section). The "box" covers the middle 50% of the sample extending from the 25th to the 75th percentile of the distribution; its length is termed the *interquartile range* (IQR). The bar inside the box is located at the median or 50th *percentile* of the sample.

A weakness of this figure is that it's hard to tell exactly what the values of the various percentiles are. A glance at the *box and whiskers plot* (Figure 1.3) made with R suggests the median of the classroom data described in Section 1.5 is about 153 cm and the interquartile range (the "box") is close to 14 cm. The minimum and maximum are located at the ends of the "whiskers."

To illustrate the use of R to create such graphs, in the next section we'll use some data I gathered while teaching mathematics and science to sixth graders.

1.4. REPORTING YOUR RESULTS

Imagine you are in the sixth grade and you have just completed measuring the heights of all your classmates.

Once the pandemonium has subsided, your instructor asks you and your team to prepare a report summarizing your results.

Actually, you have two sets of results. The first set consists of the measurements you made of you and your team members, reported in centimeters, 148.5, 150.0, and 153.0. (Kelly is the shortest incidentally, while you are the tallest.) The instructor asks you to report the minimum, the median, and the maximum height in your group. This part is easy, or at least it's easy once you look the terms up in the glossary of your textbook and discover that minimum means smallest, maximum means largest, and median is the one in the middle. Conscientiously, you write these definitions down—they could be on a test.

In your group, the minimum height is 148.5 centimeters, the median is 150.0 centimeters, and the maximum is 153.0 centimeters.

Your second assignment is more challenging. The results from all your classmates have been written on the blackboard—all 22 of them.

141, 156.5, 162, 159, 157, 143.5, 154, 158, 140, 142, 150, 148.5, 138.5, 161, 153, 145, 147, 158.5, 160.5, 167.5, 155, 137

You copy the figures neatly into your notebook computer. Using R, you store them in classdata using the command

```
classdata = c(141, 156.5, 162, 159, 157, 143.5, 154, 158,
140, 142, 150, 148.5, 138.5, 161, 153, 145, 147, 158.5,
160.5, 167.5, 155, 137)
```

Next, you brainstorm with your teammates. Nothing. Then John speaks up—he's always interrupting in class. Shouldn't we put the heights in order from smallest to largest? "Of course," says the teacher, "you should always begin by ordering your observations."

```
➤ sort(classdata)
[1] 137.0 138.5 140.0 141.0 142.0 143.5 145.0 147.0 148.5
150.0 153.0 154.0
[13] 155.0 156.5 157.0 158.0 158.5 159.0 160.5 161.0 162.0
167.5
```

Note that in R, when the resulting output takes several lines, the position of the output item in the data set is noted at the beginning of the line. Thus, 137.0 is the first item in the ordered set classdata and 155.0 is the 13th item.

"I know what the minimum is," you say (come to think of it, you are always blurting out in class, too), "137 centimeters, that's Tony."
"The maximum, 167.5, that's Pedro, he's tall," hollers someone from the back of the room.
As for the median height, the one in the middle is just 153 centimeters (or is it 154)? What does R say?

➤ median(classdata)

It is a custom among statisticians, honored by R, to report the median as the value midway between the two middle values, when the number of observations is even.

1.4.1. Picturing Data

The preceding scenario is a real one. The results reported here, especially the pandemonium, were obtained by my sixth grade homeroom at St. John's Episcopal School in Rancho Santa Marguarite, California. The problem of a metric tape measure was solved by building their own from string and a meter stick.

My students at St. John's weren't through with their assignments. It was important for them to build on and review what they'd learned in the fifth grade, so I had them draw pictures of their data. Not only is drawing a picture fun, but pictures and graphs are an essential first step toward recognizing patterns.

Begin by constructing both a strip chart and a box and whiskers plot of the classroom data using the R commands:

➤ stripchart(classdata)

and

➤ boxplot(classdata)

In S-PLUS you would type

➤ plot(classdata, 0*classdata)

All R plot commands have options that can be viewed via the R HELP menu. For example, Figure 1.3 was generated with the command

➤ boxplot(classdata, notch=TRUE, horizontal=TRUE)

Generate a strip chart and a box plot for one of the data sets you gathered in your initial assignment. Write down the values of the median, minimum, maximum, and 25th and 75th percentiles that you can infer from the box plot. Of course, you could also obtain these same values directly by using the R command, `quantile(classdata)`, which yields all the desired statistics:

0%	25%	50%	75%	100%
137.000	143.875	153.500	158.375	167.500

One word of caution: R (like most statistics software) yields an excessive number of digits. Since we only measured heights to the nearest centimeter, reporting the 25th percentile as 143.875 suggests far more precision in our measurements than actually existed. Report the value 144 centimeters instead.

A third way to depict the distribution of our data is via the histogram (Figure 1.4):

```
➢ hist(classdata)
```

To modify a histogram by increasing or decreasing the number of bars that are displayed, we make use of the "breaks" parameter as in

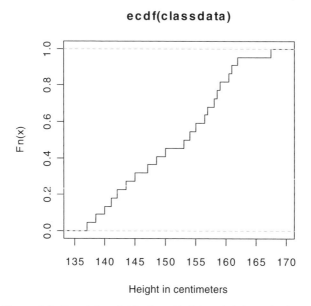

ecdf(classdata)

Height in centimeters

Figure 1.4. Cumulative distribution of heights of sixth-grade class.

➤ hist(classdata, breaks=4)

Still another way to display your data is via the *cumulative distribution function* ecdf(). To display the cumulative distribution function for the classdata, type

➤ plot(ecdf(classdata), do.points=FALSE, verticals=TRUE, xlab="Height in Centimeters")

In S-PLUS, type

➤ plotCDF(classdata, xlab = "Height in Centimeters")

Note that the x-axis of the cumulative distribution function extends from the minimum to the maximum value of your data. The y-axis reveals that the probability that a data value is less than the minimum is 0 (you knew that) and the probability that a data value is less than the maximum is 1. Using a ruler, see what x value or values correspond to 0.5 on the y scale.

Exercise 1.2. What do we call this value(s)?

Exercise 1.3. Construct histograms and cumulative distribution functions for the data you've collected.

1.4.2. Better Graphics[2]

To make your strip chart look more like the ones shown earlier, you can specify the use of a vertical line as the character to be used in plotting the points:

➤ stripchart(classdata,pch="|")

And you can create a graphic along the lines of Figure 1.2, incorporating both a box plot and strip chart, with these two commands:

➤ boxplot(classdata,horizontal=TRUE,xlab="classdata")
➤ rug(classdata)[3]

---

[2] This section and all others marked with an asterisk * expand on ways to use the R language but are not essential to an understanding of statistics.
[3] The rug() command is responsible for the tiny strip chart or rug at the bottom of the chart. Sometimes, it yields a warning message which can usually be ignored.

The first command also adds a label to the *x*-axis, giving the name of the data set, while the second command adds the strip chart to the bottom of the box plot.

1.5. TYPES OF DATA

Statistics such as the minimum, maximum, median, and percentiles make sense only if the data is *ordinal*, that is, if it can be ordered from smallest to largest. Clearly height, weight, number of voters, and blood pressure are ordinal. So are the answers to survey questions such as "How do you feel about President Bush?"

Ordinal data can be subdivided into metric and nonmetric data. *Metric* data like heights and weights can be added and subtracted. We can compute the mean as well as the median of metric data. (We can further subdivide metric data into observations like time that can be measured on a *continuous* scale and counts such as "buses per hour" that are *discrete*.)

But what is the average of "he's destroying our country" and "he's no worse than any other politician?" Such preference data is ordinal, in that it may be ordered, but it is *not* metric.

Many times, in order to analyze ordinal data, statisticians will impose a metric on it—assigning, for example, weight 1 to "Bush is destroying our country" and weight 5 to "Bush is no worse than any other politician." Such analyses are suspect, for another observer using a different set of weights might get quite a different answer.

The answers to other survey questions are not so readily ordered. For example, "What is your favorite color?" Oops, bad example, as we can associate a metric wavelength with each color. Consider instead the answers to "What is your favorite breed of dog?" or "What country do your grandparents come from?" The answers to these questions fall into nonordered categories. Pie charts and bar charts are used to display such *categorical* data and contingency tables are used to analyze it. A scatter plot of categorical data would not make sense.

Exercise 1.4. For each of the following, state whether the data is metric and ordinal, only ordinal, categorical, or you can't tell:

(a) Temperature.

(b) Concert tickets.

(c) Missing data.

(d) Postal codes.

1.5.1. Depicting Categorical Data

Three of the students in my class were of Asian origin, 18 were of European origin (if many generations back), and one was part Indian. To depict these categories in the form of a pie chart, I first entered the categorical data:

```
➢ origin = c(3,18,1)
➢ pie(origin)
```

The result looks correct, that is, *if* the data is in front of the person viewing the chart. A much more informative diagram is produced by the following R code:

```
➢ origin = c(3,18,1)
➢ names(origin) = c("Asian","European","Amerind")
➢ pie (origin, labels=names(origin))
```

All the graphics commands in R have many similar options; use R's help menu to learn exactly what these are.

A pie chart also lends itself to the depiction of ordinal data resulting from surveys. If you did a survey as your data collection project, make a pie chart of your results now.

1.6. DISPLAYING MULTIPLE VARIABLES

I'd read but didn't quite believe that one's arm span is almost exactly the same as one's height. To test this hypothesis, I had my sixth graders get out their tape measures a second time. They were to rule off the distance from the fingertips of the left hand to the fingertips of the right while the student they were measuring stood with arms outstretched like a big bird. After the assistant principal had come and gone (something about how the class was a little noisy, and though we were obviously having a good time, could we just be a little quieter), they recorded their results in the form of a two-dimensional scatter plot.

They had to reenter their height data (it had been sorted, remember), and then enter their arm span data:

```
➢ classdata = c(141, 156.5, 162, 159, 157, 143.5, 154,
  158, 140, 142, 150, 148.5, 138.5, 161, 153, 145, 147,
  158.5, 160.5, 167.5, 155, 137)
```

```
➢ armspan = c(141, 156.5, 162, 159, 158, 143.5, 155.5,
  160, 140, 142.5, 148, 148.5, 139, 160, 152.5, 142,
  146.5, 159.5, 160.5, 164, 157, 137.5)
```

This is trickier than it looks, because unless the data is entered in exactly the same order by each student in each data set, the results are meaningless. (We told you that 90% of the problems are in collecting the data and entering it in the computer for analysis. In another text of mine, *A Manager's Guide to the Design and Conduct of Clinical Trials*, I recommended eliminating paper forms completely and entering all data directly into the computer.[4]) Once the two data sets have been read in, creating a scatter plot is easy:

```
➢ height = classdata
➢ plot(height, armspan)
```

Note that we've renamed the classdata to reveal its true nature as height.
Such plots and charts have several purposes. One is to summarize the data. Another is to compare different samples or different populations (girls versus boys, my class versus your class). For example, we can enter gender data for the students, being careful to enter the gender codes in the same order in which the students' heights and arm spans already have been entered:

```
➢ sex = c("b",rep("g",7),"b",rep("g",6),rep("b",7))
```

The first student on our list is a boy, the next seven are girls, then another boy, six girls, and finally seven boys. R requires that we specify non-numeric or "character" data by surrounding the elements with quote signs. We then can use these gender data to generate side-by-side box plots of height for the boys and girls.

```
➢ sexf = factor(sex)
➢ plot(sexf,height)
```

The R function **factor** () tells the computer to treat gender as a categorical variable, one that in this case takes two values "b" and "g." The **plot()** function will not work until character data has been converted to factors.
The primary value of charts and graphs is as aids to critical thinking. The figures in this specific example may make you start wondering about the

---

[4] We'll discuss how to read data files using R later in this chapter.

uneven way adolescents go about their growth. The exciting thing, whether you are a parent or a middle-school teacher, is to observe how adolescents get more heterogeneous, more individual with each passing year.

Exercise 1.5. Use the preceding R code to display and examine the indicated charts for my classroom data.

Exercise 1.6. Modify the preceding R code to obtain side-by-side box plots for the data you've collected.

1.6.1. From Observations to Questions

You may want to formulate your theories and suspicions in the form of questions, too: Are girls in the sixth grade taller on average than sixth-grade boys (not just those in my sixth-grade class, but in all sixth-grade classes)? Are they more homogeneous, that is, less variable, in terms of height? What is the average height of a sixth grader? How reliable is this estimate? Can height be used to predict arm span in sixth grade? Can it be used to predict the arm spans of students of any age?

You'll find straightforward techniques in subsequent chapters for answering these and other questions. First, we suspect, you'd like the answer to one really big question: Is statistics really much more difficult than the sixth-grade exercise we just completed? No, this is about as complicated as it gets.

1.7. MEASURES OF LOCATION

Far too often, we find ourselves put on the spot, forced to come up with a one-word description of our results when several pages or, better still, several charts would do. "Take all the time you like," coming from a boss, usually means, "Tell me in ten words or less."

If you were asked to use a single number to describe data you've collected, what number would you use? One answer is "the one in the middle," the *median* that we defined earlier in this chapter.

In the majority of cases, we recommend using the *arithmetic mean* rather than the median. To calculate the mean of a sample of observations by hand, one adds up the values of the observations, then divides by the number of observations in the sample. If we observe 3.1, 4.5, and 4.4, the arithmetic mean would be 12/3 = 4. In symbols, we write the mean of a sample of n observations, X_i with $i = 1, 2, \ldots, n$, as[5]

[5] The Greek letter Σ is pronounced sigma.

$$(X_1 + X_2 + \cdots + X_n)\Big/ n = \frac{1}{n}\sum_{i=1}^{n} X_i = \overline{X}.$$

Is adding a set of numbers and then dividing by the number in the set too much work? To find the mean height of the students in Good's classroom, use R and enter

➢ mean(height)

A playground seesaw (or teeter-totter) is symmetric in the absence of kids. Its midpoint or median corresponds to its center of gravity or its mean. If you put a heavy kid at one end and two light kids at the other so that the seesaw balances, the mean will still be at the pivot point, but the median is located at the second kid.

Another population parameter of interest is the most frequent observation or *mode*. In the sample 2, 2, 3, 4, and 5, the mode is 2. Often the mode is the same as the median or close to it. Sometimes it's quite different, and sometimes, particularly when there is a mixture of populations, there may be several modes.

Consider the data on heights collected in my sixth-grade classroom. The mode is at 157.5 cm. But aren't there really two modes, one corresponding to the boys, the other to the girls in the class? As you can see on typing the command

➢ hist(classdata, xlab="Heights of Students in Dr.Good's
  Class (cm)")

a histogram of the heights provides evidence of two modes (Figure 1.5). When we don't know in advance how many subpopulations there are, modes serve a second purpose: to help establish the number of subpopulations.

Exercise 1.7. Compare the mean, median, and mode of the data you've collected.

Exercise 1.8. A histogram can be of value in locating the modes when there are 20 to several hundred observations, because it groups the data. Use R to draw histograms for the data you've collected.

1.7.1. Which Measure of Location?

The means (arithmetic and geometric), the median, and the mode are examples of sample *statistics*. Statistics serve three purposes:

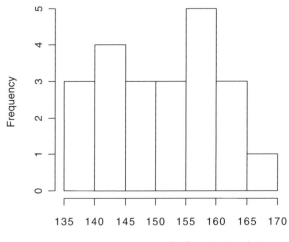

Figure 1.5. Histogram.

1. Summarizing data.
2. Estimating population parameters.
3. Aids to decision making.

Our choice of one statistic rather than another depends on the use(s) to which it is to be put.

THE CENTER OF A POPULATION

Median: the value in the middle; the halfway point; that value which has equal numbers of larger and smaller elements around it.

Arithmetic mean or arithmetic average: the sum of all the elements divided by their number or, equivalently, that value such that the sum of the deviations of all the elements from it is zero.

Mode: the most frequent value. If a population consists of several sub-populations, there may be several modes.

For summarizing data, graphs—box plots, strip plots, cumulative distribution functions, and histograms—are essential. If you're not going to use a histogram, then for samples of 20 or more be sure to report the number of modes.

We always recommend using the median in two instances:

1. If the data are ordinal but not metric.
2. When the distribution of values is highly skewed with a few very large or very small values.

Two good examples are incomes and house prices. A recent *Los Angeles Times* ad featured a great house in Beverly Hills at $80 million US. A house like that has a large effect on the mean price of homes in an area. The median house price is far more representative than the mean, even in Beverly Hills.

The weakness of the arithmetic mean is that it is too easily biased by extreme values. If we eliminate Pedro from our sample of sixth graders—he's exceptionally tall for his age at 5′7″—the mean would change from 151.6 to 3167/21 = 150.8 cm. The median would change to a much lesser degree, shifting from 153.5 to 153 cm. Because the median is not as readily biased by extreme values, we say that the median is more *robust* than the mean.

The *geometric mean* is the appropriate measure of location when we are expressing changes in percentages, rather than absolute values. For example, if in successive months the cost of living was 110%, 105%, 110%, 115%, 118%, 120%, and 115% of the value in the base month, the geometric mean would be $(1.1 \times 1.05 \times 1.1 \times 1.15 \times 1.18 \times 1.2 \times 1.15)^{1/7}$.

We can obtain the geometric mean directly from its definition

➢ `(1.2*2.3*4.0*3*5.1)∧(1/5)`

Here, "∧(x)" is read by R as "raise to the *x*-th power." As the data is already in a vector named ourdata, we can get the same result from

➢ `prod(ourdata)∧(1/5)`

The R function **prod(v)** computes the product of the elements making up a vector **v**. Finally, if we have forgotten how many observations there are in ourdata, we can ask R to count them before computing the geometric mean:

➢ `prod(ourdata)∧(1/length(ourdata))`

Note: The names of R functions like **prod()** and **mean()** must be spelled exactly as they appear. The names of variables like "ourdata" and "nudata" are made-up names and you are free to use any other made-up names you feel more comfortable with. But if you use the spelling "nudata" in one part of your program, don't expect R to recognize the name spelled "newdata" in another.

For Estimation. In deciding which *sample statistic* to use in estimating the corresponding *population parameter,* we need to distinguish between precision and accuracy. Let us suppose Robin Hood and the Sheriff of Nottingham engage in an archery contest. Each is to launch three arrows at a target 50 meters (half a soccer pitch) away. The Sheriff launches first and his three arrows land one atop the other in a dazzling display of shooting *precision.* Unfortunately, all three arrows penetrate and fatally wound a cow grazing peacefully in the grass nearby. The Sheriff's *accuracy* leaves much to be desired.

We can show mathematically that for very large samples the sample median and the population median will almost coincide. The same is true for large samples and the mean. Alas, "large" may mean larger than we can afford. With small samples, the accuracy of an estimator is always suspect.

With most of the samples we encounter in practice, we can expect the value of the sample median and virtually any other estimator to vary from sample to sample. One way to find out for small samples how *precise* a method of estimation is would be to take a second sample the same size as the first and see how the estimator varies between the two. Then a third, and fourth, . . . , say, 20 samples. *But a large sample will **always** yield more precise results than a small one.* So, if we'd been able to afford it, the sensible thing would have been to take 20 times as large a sample to begin with.[6]

Still, there is an alternative. We can treat our sample as if it were the original population and take a series of *bootstrap samples* from it. The variation in the value of the estimator from bootstrap sample to bootstrap sample will be a measure of the variation to be expected in the estimator had we been able to afford to take a series of samples from the population itself. The larger the size of the original sample, the closer it will be in composition to the population from which it was drawn, and the more accurate this measure of precision will be.

---

[6] Of course, there is a point at which each additional observation will cost more than it yields in information. The bootstrap described here will also help us to find the "optimal" sample size.

Exercise 1.9. Are women gabbier then men? Perhaps, not when it comes to email attachments. Compute the mean and the median of attachment size seperatly for each sex using he following data. Which measure of location do you feel is more appropiate in this example:

```
emailsize = c(14,91,5,82,38,1000,1,1,44,1,379,91,18,26)
emailgender = c("male", "male", "female", rep("male", 6),
"female", rep("male", 3), "female")
```

1.7.2. The Bootstrap

Let's see how this process, called bootstrapping, would work with a specific set of data. Once again, here are the heights of the 22 students in Dr. Good's sixth-grade class, measured in centimeters and ordered from shortest to tallest:

137.0	138.5	140.0	141.0	142.0	143.5	145.0	147.0	148.5
150.0	153.0	154.0	155.0	156.5	157.0	158.0	158.5	159.0
160.5	161.0	162.0	167.5					

Let's assume we record each student's height on an index card, 22 index cards in all. We put the cards in a big hat, shake them up, pull one out and make a note of the height recorded on it. We *return the card to the hat* and repeat the procedure for a total of 22 times till I have a second sample, the same size as the original. Note that we may draw Jane's card several times as a result of using this method of *sampling with replacement.*

Our first bootstrap sample, arranged in increasing order of magnitude for ease in reading, might look like this:

138.5	138.5	140.0	141.0	141.0	143.5	145.0	147.0	148.5
150.0	153.0	154.0	155.0	156.5	157.0	158.5	159.0	159.0
159.0	160.5	161.0	162					

Several of the values have been repeated; not surprising as we are sampling with replacement, treating the original sample as a stand-in for the much larger population from which the original sample was drawn. The minimum of this bootstrap sample is 138.5, higher than that of the original sample; the maximum at 162.0 is less than the original, while the median remains unchanged at 153.5.

137.0	138.5	138.5	141.0	141.0	142.0	143.5	145.0	145.0
147.0	148.5	148.5	150.0	150.0	153.0	155.0	158.0	158.5
160.5	160.5	161.0	167.5					

142.25   Medians of Bootstrap Samples   158.25

Figure 1.6. One-way stripplot.

In this second bootstrap sample, again we find repeated values—quick, what are they? This time the minimum, maximum, and median are 137.0, 167.5, and 148.5, respectively.

Two bootstrap samples cannot tell us very much. But suppose we were to take 50 or a hundred such samples. Figure 1.6 is a one-way strip plot of the medians of 50 bootstrap samples taken from the classroom data. These values provide an insight into what might have been had we sampled repeatedly from the original population.

Quick question: What is that population? Does it consist of all classes at the school where I was teaching? All sixth-grade classes in the district? All sixth-grade classes in the state? The school was Episcopalian, so perhaps the population was all sixth-grade classes in Episcopalian schools.

Exercise 1.10. Our original question, you'll recall, is: Which is the least variable (most precise) estimate—mean or median? To answer this question, at least for several samples, let us apply the bootstrap, first to our classroom data and then to the data we collected in Exercise 1.1. You'll need the following R listing:

```
➢ #Comments to R code all start with the pound sign #
➢ #This program selects 100 bootstrap samples from your data
➢ #and then produces a boxplot of the results.
➢ #first, we give a name, urdata, to the observations in our
   original sample
➢ urdata = c( , . . .)
➢ #Record group sizes
➢ n = length(urdata)
➢ #set number of bootstrap samples
➢ N =100
➢ stat = numeric(N)    #create a vector in which to store the
                        results
➢                #the elements of the vector will be numbered
                  from 1 to N
➢ #Set up a loop to generate a series of bootstrap samples
➢ for (i in 1:N){
+ #bootstrap sample counterparts to observed samples are
  denoted with "B"
+ urdataB= sample (urdata, n, replace=T)
+ stat[i] = mean(urdataB)
+ }
➢ boxplot (stat)
```

1.8. SAMPLES AND POPULATIONS

If it weren't for person-to-person variation, it really would be easy to find out what brand of breakfast cereal people prefer or which movie star they want as their leader. Interrogate the first person you encounter on the street and all will be revealed. As things stand, we must either pay for and take a total census of everyone's view (the cost of the 2003 recall election in California pushed an already near-bankrupt state one step closer to the edge) or take a sample and learn how to extrapolate from that sample to the entire population.

In each of the data collection examples in Section 1.2, our observations were limited to a sample from a population. We measured the height, circumference, and weight of a dozen humans (or dogs, or hamsters, or frogs, or crickets) but not all humans or dogs or hamsters. We timed *some* individuals (or frogs or turtles) in races but not *all*. We interviewed some fellow students but not all.

If we had interviewed a different set of students, would we have gotten the same results? Probably not. Would the means, medians, interquartile ranges (IQRs), and so forth have been similar for the two sets of students? Maybe, *if* the two samples had been large enough and similar to each other in composition.

If we interviewed a sample of women and a sample of men regarding their views on women's right to choose, would we get similar answers? Probably not, as these samples were drawn from completely different populations (different, that is, with regard to their views on women's right to choose). If we want to know how the citizenry as a whole feels about an issue, we need to be sure to interview both men and women.

In every statistical study, two questions immediately arise:

1. How large should my sample be?
2. How can I be sure this sample is representative of the population in which my interest lies?

By the end of Chapter 5, we'll have enough statistical knowledge to address the first question, but we can start now to discuss the second.

After I deposited my ballot in a recent election, I walked up to the interviewer from the *Los Angeles Times* who was taking an exit poll and offered to tell her how I'd voted. "Sorry," she said, "I can only interview every ninth person."

What kind of a survey wouldn't want my views? Obviously, a survey that wanted to ensure that shy people were as well represented as boisterous and that a small group of activists couldn't bias the results.[7]

---

[7] To see how surveys could be biased deliberately, you might enjoy reading Grisham's *The Chamber*.

One sample we would all insist be representative is the jury.[8] The Federal Jury Selection and Service Act of 1968 as revised[9] states that citizens cannot be disqualified from jury duty "on account of race, color, religion, sex, national origin or economic status."[10] The California Code of Civil Procedure, Section 197, tells us *how* to get a representative sample. First, you must be sure your sample is taken from the appropriate population. In California's case, the "list of registered voters and the Department of Motor Vehicles list of licensed drivers and identification card holders . . . shall be considered inclusive of a representative cross section of the population." The Code goes on to describe how a table of random numbers or a computer could be used to make the actual selection. The bottom line is that to obtain a random, representative sample:

- Each individual (or item) in the population must have an equal probability of being selected.
- No individual (item) or class of individuals may be discriminated against.

There's good news and bad news. The bad news is that any individual sample may not be representative. You can flip a coin six times and every so often it will come up heads six times in a row. A jury may consist entirely of white males. The good news is that as we draw larger and larger samples, the samples will resemble more and more closely the population from which they are drawn.

Exercise 1.11. For each of the three data collection examples of Section 1.2, describe the populations you would hope to extend your conclusions to and how you would go about ensuring that your samples were representative in each instance.

1.8.1. Drawing a Random Sample

Recently, one of our clients asked for help with an audit. Some errors had been discovered in an invoice they'd submitted to the government for reimbursement. Since this client, an HMO, made hundreds of such submissions each month, they wanted to know how prevalent such errors were. Could we help them select a sample for analysis?

---

[8] Unless of course, we are the ones on trial.
[9] 28 U.S.C.A. x1861 et seq. (1993).
[10] See 28 U.S.C.A. x1862 (1993).

We could, but we needed to ask the client some questions first. We had to determine what the population was from which the sample would be taken and what constituted a *sampling unit.*

Were we interested in all submissions or just some of them? The client told us that some submissions went to state agencies and some to federal, but for audit purposes their sole interest was in certain federal submissions, specifically in submissions for reimbursement for a certain type of equipment. Here, too, a distinction needed to be made between custom equipment (with respect to which there was virtually never an error) and more common off-the-shelf supplies. At this point in the investigation, our client breathed a sigh of relief. We'd earned our fee, it appeared, merely by observing that instead of 10,000 plus potentially erroneous claims, the entire population of interest consisted of only 900 or so items.

(When you read earlier that 90% of the effort in statistics was in collecting the data, we meant exactly that.)

Our client's staff, like that of most businesses, was used to working with an electronic spreadsheet. "Can you get us a list of all the files in spreadsheet form?" we asked.

They could and did. The first column of the spreadsheet held each claim's ID. The second held the date. We used the spreadsheet's sort function to sort all the claims by date, and then deleted all those that fell outside the date range of interest. Next , a new column was inserted and in the top cell (just below the label row) of the new column, we put the command `rand()`. We copied this command all the way down the column.

A series of numbers between 0 and 1 was displayed down the column. To lock these in place, we went to the Tools menu, clicked on "options," and then on the calculation tab. Next, we made sure that Calculation was set to manual and there was no check mark opposite "recalculate before save."

Now, we resorted the data based on the results of this column. Beforehand, we'd decided there would be exactly 35 claims in the sample, so we simply cut and pasted the top 35 items.

1.8.2. Using R to Draw a Sample*

The easy part of using R to draw a sample is the part where the data we want is already in the computer memory and is stored in the vector "DATA." If we want 35 observations from DATA selected at random, we would just need to write

```
➢ sample (DATA, 35)
```

Now suppose we have our data stored in a file called class.dat on a hard disk or CD and that this file has the following form:

1. The first line of class.dat has a name for each variable in the data set.
2. Each additional line or record in class.dat has as its first item a row label followed by the values for each variable in the record, separated by blanks.

Here is an example:

	Name	Age	Gender	Height	ArmSpan
01	Pedro	11	B	167	163
02	Renee	11	G	163	162.5
03	Bill	10	B	153	152

We could bring class.dat into our computer's memory and place it for use in an R frame with the command

```
➢ classdat = read.table("class.dat")
```

If class.dat was on a hard disk in the file folder \mywork or /mywork, we would store it for use in an R frame with the command

```
➢ classdat = read.table("/mywork/class.dat")
```

(*Note*: Use this form of referring to a file folder no matter whether you are working in a Windows, UNIX, or Mac environment.)

If commas rather than blanks separate the data items in each record of your file, use the R command

```
➢ classdat = read.table("/mywork/class.dat", sep=",")
```

Before we may use the classdat variable names, we must first attach the data frame with the command

```
➢ attach(classdat)
```

Then we may execute R commands such as

```
➢ mean (Height)
➢ boxplot(ArmSpan)
```

Note that R commands are case sensitive. The following instructions won't work:

```
➢ mean (height)
➢ boxplot(Armspan)
```

To select a sample of ten random student records from classdat, use the commands

```
➢ nn = nrow(classdat)
➢ rr = sample(1:nn,10)
➢ samp = classdat[rr,]
```

or, put the three commands together into one.

```
➢ samp= classdat [sample(1:nrow(classdat),10),]
```

Now, let's put all we've learned together in one program:

```
➢ classdat = read.table("/mywork/class.dat")
➢ attach (classdat)
➢ samp= classdat [sample(1:nrow(classdat),10),]
➢ mean (Height)
[1] 161.2
➢ boxplot(ArmSpan)
```

1.8.3. Ensuring the Sample Is Representative

Exercise 1.12. We've already noted that a random sample might not be representative. By chance alone, our sample might include men only, or African-Americans but no Asians, or no smokers. How would you go about ensuring that a random sample is representative?

1.9. VARIATION—WITHIN AND BETWEEN

Our work so far has revealed that the values of our observations vary within a sample as well as between samples taken from the same population. Not surprisingly, we can expect even greater variability when our samples are drawn from different populations. Several different statistics are used to characterize and report on the *within-sample variation.*

The most common is termed the *variance* and is defined as the sum of the squares of the deviations of the individual observations about their mean divided by the sample size minus 1. In symbols, if our observations are labeled X_1, X_2, up to X_n, and the mean of these observations is written as $\overline{X}$, then the variance σ^2 (pronounced sigma squared) is equal to $[1/(n-1)] \Sigma_{i=1}^{n} (X_i - \overline{X})^2$.

Exercise 1.13. What is the sum of the deviations of the observations from their arithmetic mean? That is, what is $\Sigma_{i=1}^{n} (X_i - \overline{X})$?

The problem with using the variance is that if our observations, on temperature, for example, are in degrees Celsius, then the variance would be expressed in square degrees, whatever these are. More often, we report the *standard deviation* σ, the square root of the variance, as it is in the same units as our observations.

Reporting the standard deviation has the further value that if our observations come from a *normal* distribution like that depicted in Figure 1.7, then we know that the probability is 68% that an observation taken from such a population lies within plus or minus one standard deviation of the population mean.

If we have two samples and aren't sure if they come from the same population, one way to check is to express the difference in the sample means, the *between-sample variation*, in terms of the within-sample variation or standard deviation. We'll investigate this approach in Chapter 3.

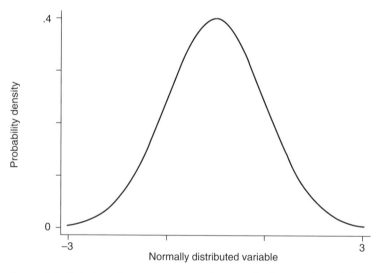

Figure 1.7. Bell-shaped symmetric curve of a normally distributed population.

If the observations do not come from a normal distribution, then the standard deviation is less valuable. In such a case, we might want to report as a measure of dispersion the sample *range,* which is just the maximum minus the minimum, or the *interquartile range,* which is the distance between the 75th and 25th percentiles. From a box plot of our data, we can get eyeball estimates of the range as the distance from whisker end to whisker end, and the interquartile range, which is the length of the box. Of course, to obtain exact values, we would use R's quantile function.

Exercise 1.14. What are the variance, standard deviation, and interquartile range of the classroom data? What are the 90th and 5th percentiles? (To do this exercise, you'll need the R functions var, sqrt, and quantile. If you're not sure how to use them, use the R HELP menu.)

This next exercise is only for those familiar with calculus.

Exercise 1.15. Show that we can minimize the sum of squares $\sum_{i=1}^{n}(X_i - A)^2$ if we let A be the sample mean.

1.10. SUMMARY AND REVIEW

In this chapter, you learned R's syntax and a number of R and S-PLUS commands with which to:

- Perform mathematical functions (prod, sqrt)
- Create graphs (box plot, hist, plot, pie, strip chart)
- Compute statistics (mean, median, quantile, variance)
- Manipulate data and create vectors of observations (sort, numeric, factor, name)
- Control program flow (for)
- Select random samples (sample)
- Read data from tables (read.table)

The best way to summarize and review the statistical material we've covered so far is with the aid of three additional exercises.

Exercise 1.16. Make a list of all the *italicized* terms in this chapter. Provide a definition for each one along with an example.

Exercise 1.17. The following data on the relationship of performance on the LSATs to GPA is drawn from a population of 82 law schools. We'll look at this data again in Chapters 3 and 4.

LSAT = c(576, 635, 558, 578, 666, 580, 555, 661, 651, 605, 653, 575, 545, 572, 594)

GPA = c(3.39, 3.3, 2.81, 3.03, 3.44, 3.07, 3, 3.43, 3.36, 3.13, 3.12, 2.74, 2.76, 2.88, 2.96)

Make box plots and histograms for both the LSAT score and GPA. Tabulate the mean, median, interquartile range, standard deviation, and 95th and 5th percentiles for both variables.

Exercise 1.18. I have a theory that literally all aspects of our behavior are determined by our birth order (oldest/only, middle, youngest) including clothing, choice of occupation, and sexual behavior. How would you go about collecting data to prove or disprove some aspect of this theory?

2

PROBABILITY

In this chapter, you'll learn the rules of probability and apply them to games of chance, jury selection, surveys, and blood types. You'll use R to generate simulated random data and learn how to create your own R functions.

2.1. PROBABILITY

Because of the variation inherent in the processes we study, we are forced to speak in probabilistic terms rather than absolutes. We talk about the probability that a sixth-grader is exactly 150 cm tall or, more often, that the height will lie between two values such as 150 cm and 155 cm. The events we study may happen a large proportion of the time, or "almost always," but seldom "always" or "never."

Rather arbitrarily, and some time ago, it was decided that probabilities would be assigned a value between 0 and 1, that events that were certain to occur would be assigned probability 1, and that events that would "never" occur would be given probability 0. When talking about a set of *equally likely* events, such as the probability that a fair coin will come up heads, or an unweighted die will display a "6," this limitation makes a great deal of sense. A coin has two sides; we say the probability it comes up heads is a half and

---

Introduction to Statistics Through Resampling Methods and R/S-PLUS®, By Phillip I. Good
Copyright © 2005 by John Wiley & Sons, Inc.

the probability of tails is a half also: $\frac{1}{2}+\frac{1}{2}=1$, the probability that a coin comes up something.[1] Similarly, the probability that a six-sided die displays a "6" is 1/6. The probability it does *not* display a 6 is $1 - 1/6 = 5/6$.

For every dollar you bet, roulette wheels pay off $36 if you win. This certainly seems fair, until you notice that not only does the wheel have slots for the numbers 1 through 36, but there is a slot for 0, and sometimes for double 0, and for triple 000 as well. Thus the real probabilities of winning and losing are, respectively, 1 chance in 39 and 38/39. In the long run, you lose one dollar thirty-eight times as often as you win $36. Even when you win, the casino pockets your dollar, so that in the long run the casino pockets $3 for every $39 that is bet. (And from whose pockets does that money come?)

Ahh, but you have a clever strategy called a *martingale*. Every time you lose, you simply double your bet. So if you lose a dollar the first time, you lose two dollars the next. Hmm. Since the casino always has more money than you do, you still end up broke. Tell me again why this is a clever strategy.

Exercise 2.1. List the possible ways in which the following can occur:

(a) A person, call him Bill, is born on a specific day of the week.

(b) Bill and Alice are born on the same day of the week.

(c) Bill and Alice are born on different days of the week.

(d) Bill and Alice play a round of a game called "paper, scissor, stone" and simultaneously display either an open hand, two fingers, or a closed fist.

Exercise 2.2. Match the probabilities with their descriptions. A description may match more than one probability.

(a) −1	(1) infrequent
(b) 0	(2) virtually impossible
(c) 0.10	(3) certain to happen
(d) 0.25 inch	(4) typographical error
(e) 0.50	(5) more likely than not
(f) 0.80	(6) certain
(g) 1.0	(7) highly unlikely
(h) 1.5	(8) even odds
	(9) highly likely

---

[1] I had a professor at Berkeley who wrote a great many scholarly articles on the subject of "coins that stand on edge," but then that is what professors at Berkeley do.

To determine whether a gambling strategy or a statistic is optimal, we need to know a few of the laws of probability. These laws show us how to determine the probabilities of combinations of events. For example, if the probability that an event A will occur is $P\{A\}$, then the probability that A won't occur $P\{\text{not } A\} = 1 - P\{A\}$. This makes sense since either the event A occurs or it doesn't, and thus $P\{A\} + P\{\text{not } A\} = 1$.

We'll also be concerned with the probability that both A and B occur, $P\{A \text{ and } B\}$, or with the probability that either A occurs or B occurs or both do, $P\{A \text{ or } B\}$. If two events A and B are *mutually exclusive*, that is, if when one occurs the other cannot possibly occur, then the probability that A **or** B will occur, $P\{A \text{ or } B\}$, is the sum of their separate probabilities. (Quick, what is the probability that both A **and** B occur?) The probability that a six-sided die will show an odd number is thus $\frac{3}{6}$ or $\frac{1}{2}$. The probability that a six-sided die will *not* show an even number is equal to the probability that a six-sided die will show an odd number.

2.1.1. Events and Outcomes

An *outcome* is something we can observe. For example, "the coin lands heads" or "an odd number appears on the die." Outcomes are made up of *events* that may or may not be completely observable. The referee tosses the coin into the air, it flips over three times before he catches it and places it face upward on his opposite wrist. "Heads," and Manchester United gets the call. But the coin might also have come up heads had the coin been tossed higher in the air so that it spun three and a half or four times before being caught. A literal infinity of events makes up the single observed outcome, "Heads."

The outcome "an odd number appears on the six-sided die" is composed of three outcomes—1, 3, and 5—each of which can be the result of any of an infinity of events. By definition, events are mutually exclusive. Outcomes may or may not be mutually exclusive, depending on how we aggregate events.

2.1.2. Venn Diagrams

An excellent way to gain insight into the distinction between events and outcomes and the laws of probability is via the Venn diagram.[2] Figure 2.1 pictures two overlapping outcomes, A and B. For example, A might consist

---

[2] Curiously, not a single Venn diagram is to be found in John Venn's text, *The Logic of Chance*, published by Macmillan and Co., London, 1866, with a third edition in 1888.

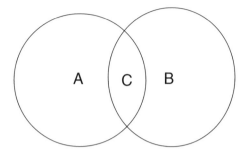

Figure 2.1. Venn diagram depicting two overlapping outcomes.

of all those who respond to a survey that they are nonsmokers, while B corresponds to the outcome that the respondent has lung cancer.

Every point in the figure corresponds to an event. The events within the circle A all lead to the outcome A. Note that many of the events or points in the diagram lie outside both circles. These events correspond to the outcome, "neither A nor B" or, in our example, "an individual who does smoke and does not have lung cancer."

The circles overlap; thus outcomes A and B are not mutually exclusive. Indeed, any point in the region of overlap between the two, marked C, leads to the outcome "A and B." What can we say about individuals who lie in region C?

Exercise 2.3. Construct a Venn diagram corresponding to the possible outcomes of throwing a six-sided die. (I find it easier to use squares than circles to represent the outcomes, but the choice is up to you.) Does every event belong to one of the outcomes? Can an event belong to more than one of these outcomes? Now, shade the area that contains the outcome "the number face up on the die is odd." Use a different shading to outline the outcome "the number on the die is greater than 3."

Exercise 2.4. Are the outcomes "the number face up on the die is odd" and "the number on the die is greater than 3" mutually exclusive?

You'll find many excellent Venn diagrams illustrating probability concepts at `http://stat-www.berkeley.edu/~stark/Java/Venn.htm`.

Exercise 2.5. According to the *Los Angeles Times*, scientists are pretty sure planetoid Sedna has a moon, though as of April 2004 they'd been unable to see it. The scientists felt at the time there was a 1 in 100 possibility that the moon might have been directly in front of or behind the planetoid when they looked for it, and a 5 in 100 possibility that they'd misinterpreted Sedna's rotation rate. How do you think they came up with those probabilities?

IN THE LONG RUN: SOME MISCONCEPTIONS

When events occur as a result of chance alone, anything can happen and usually will. You roll craps seven times in a row, or you flip a coin ten times and ten times it comes up heads. Both these events are unlikely, but they are not impossible. Before reading the balance of this section, test yourself by seeing if you can answer the following:

You've been studying a certain roulette wheel that is divided into 38 sections for over four hours now and not once during those four hours of continuous play has the ball fallen into the number 6 slot. Which of the following do you feel is more likely?

(a) Number 6 is bound to come up soon.
(b) The wheel is fixed so that number 6 will never come up.
(c) The odds are exactly what they've always been and in the next four hours number 6 will probably come up about $\frac{1}{38}$th of the time.

If you answered (b) or (c) you're on the right track. If you answered (a), think about the following equivalent question: You've been studying a series of patients treated with a new experimental drug all of whom died in excruciating agony despite the treatment. Do you conclude the drug is bound to cure somebody sooner or later and take it yourself when you come down with the symptoms? Or do you decide to abandon this drug and look for an alternative?

2.2. BINOMIAL

Many of our observations take a yes/no or dichotomous form: "My headache did/didn't get better." "Chicago beat/was beaten by Los Angeles." "The respondent said she would/wouldn't vote for Dean." The simplest example of a *binomial trial* is that of a coin flip: heads I win, tails you lose.

If the coin is fair, that is, if the only difference between the two mutually exclusive outcomes lies in their names, then the probability of throwing a head is 1/2, and the probability of throwing a tail is also 1/2. (That's what I like about my bet, either way I win.)

By definition, the probability that something will happen is 1, the probability that nothing will occur is 0. All other probabilities are somewhere in between.[3]

What about the probability of throwing heads twice in a row? Ten times in a row? If the coin is fair and the throws independent of one another, the answers are easy: $\frac{1}{4}$ and $\frac{1}{1024}$ or $(\frac{1}{2})^{10}$.

These answers are based on our belief that when the only differences among several possible *mutually exclusive* outcomes are their labels, "heads" and "tails," for example, the various outcomes will be *equally likely*. If we flip two fair coins or one fair coin twice in a row, there are four possible outcomes HH, HT, TH, and TT. Each outcome has equal probability of occurring. The probability of observing the one outcome in which we are interested is 1 in 4 or $\frac{1}{4}$. Flip the coin ten times and there are 2^{10} or a thousand possible outcomes; one such outcome might be described as HTTTTTTTTH.

Unscrupulous gamblers have weighted coins so that heads comes up more often than tails. In such a case, there is a real difference between the two sides of the coin and the probabilities will be different from those described above. Suppose as a result of weighting the coin, the probability of getting a head is now p, where $0 \leq p \leq 1$, and the complementary probability of getting a tail (or not getting a head) is $1 - p$, because $p + (1 - p) = 1$. Again, we ask the question: What is the probability of getting two heads in a row? The answer is p^2. Here is why.

To get two heads in a row, we must throw a head on the first toss, which we expect to do in a proportion p of attempts. Of this proportion, only a further fraction p of two successive tosses also end with a head, that is, only $p * p$ trials result in HH. Similarly, the probability of throwing ten heads in a row is p^{10}.

By the same line of reasoning, we can show the probability of throwing nine heads in a row followed by a tail when we use the same weighted coin each time is $p^9(1 - p)$. What is the probability of throwing 9 heads in 10 trials? Is it also $p^9(1 - p)$? No, for the outcome "nine heads out of ten" includes the case where the first trial is a tail and all the rest are heads, the second trial is a tail and all the rest are heads, the third trial is . . . , and so forth. Ten different ways in all. These different ways are *mutually exclusive*; that is, if one of these events occurs, the others are excluded. The probability of the overall event is the sum of the individual probabilities or 10 $p^9(1 - p)$.

---

[3] If you want to be precise, the probability of throwing a head is probably only 0.49999, and the probability of a tail is also only 0.49999. The leftover probability of 0.00002 is the probability of all the other outcomes—the coin stands on edge, a sea gull drops down out of the sky and takes off with it, and so forth.

Exercise 2.6. What is the probability that if you flipped a fair coin you would get heads five times in a row?

Exercise 2.7. The strength of support for our candidate seems to depend on whether we are interviewing men or women: 50% of male voters support our candidate, but only 30% of women. What percentage of women favor some other candidate? If we select a woman and a man at random and ask which candidate they support, in what percentage of cases do you think both will say they support our candidate?

Exercise 2.8. Would your answer to the previous question be the same if the man and the woman were co-workers?

Exercise 2.9. Which do you think would be preferable in a customer-satisfaction survey? To ask customers if they were or were not satisfied? Or to ask them to specify their degree of satisfaction on a five-point scale? Why?

2.2.1. Permutations and Rearrangements

What is the probability of throwing exactly five heads in ten tosses of a coin? The answer to this last question requires we understand something about permutations and combinations or rearrangements, a concept that will be extremely important in succeeding chapters.

Suppose we have three horses in a race. Call them A, B, and C. A could come in first, B could come in second, and C would be last. ABC is one possible outcome or permutation. But so are ACB, BAC, BCA, CAB, and CBA—six possibilities or *permutations* in all. Now suppose we have a nine-horse race. We could write down all the possibilities, or we could use the following trick: We choose a winner (nine possibilities); we choose a second place finisher (eight remaining possibilities), and so forth until all positions are assigned. A total of $9! = 9 \times 8 \times 7 \times 6 \times 5 \times 4 \times 3 \times 2 \times 1$ possibilities in all. Had there been N horses in the race, there would have been $N!$ possibilities. $N!$ is read "N factorial."

Note that $N! = N(N - 1)!$

Normally in a horse race, all our attention is focused on the first three finishers. How many possibilities are there? Using the same reasoning, it is easy to see there are $9 \times 8 \times 7$ possibilities or $9!/6!$. Had there been N horses in the race, there would have been $N!/(N - 3)!$ possibilities.

Suppose we ask a slightly different question: In how many different ways can we select three horses from nine entries without regard to order (i.e., we don't care which comes first, which second, or which third)? In the previous example, we distinguished between first, second, and third place finishers; now we're saying the order of finish doesn't make any difference. We already know there are $3! = 3 \times 2 \times 1 = 6$ different permutations of the three horses that finish in the first three places. So we take our answer to the preceding question 9!/6! and divide this answer in turn by 3!. We write the result as $\binom{9}{3}$ which is usually read as 9 choose 3. Note that

$$\binom{9}{3} = \binom{9}{6}$$

In how many different ways can we assign nine cell cultures to two unequal experimental groups, one with three cultures and one with six? This would be the case if we had nine labels and three of the labels read "vitamin E" while six read "control." If we could distinguish the individual labels, we could assign them in 9! different ways. But the order they are assigned in each of the experimental groups, 3! ways in the first instance and 6! in the other, won't affect the results. Thus there are only 9!/6!3! or

$$\binom{9}{3} = 84$$

distinguishable ways. We can generalize this result to show the number of distinguishable ways N items can be assigned to two groups, one of k items, and one of $N - k$ is

$$\frac{N!}{k!(N-k)!} = \binom{N}{K}$$

What if we were to divide these same nine cultures among three equal-sized experimental groups? Then we would have 9!/3!3!3! distinguishable ways or *rearrangements*, written as

$$\binom{9}{3 \quad 3 \quad 3}$$

Exercise 2.10. What is the value of 4!?

Exercise 2.11. In how may different ways can we divide eight subjects into two equal-sized groups?

*Programming Your Own Functions in R.** To tell whether our answer to question 2.2 is correct, we can program the factorial function in R. We make use of our knowledge that $N! = N(N - 1)!$:

```
➢ fact   = function (num)
+ {
+ if (num==0) return (1)
+ else return (num*fact(num-1))
+ }
```

or, we could obtain the answer directly with the listing

```
➢ fact   = function (num) {
+ prod(1:num)
+ }
```

Here, `1:num` yields a sequence of numbers from `1` to `num` and **prod()** gives the product of these numbers.

In S-PLUS you also can use the built-in **factorial()** function.

To check your answer to the exercise, just type **fact**(4) after the ➢. Note that this function only makes sense for positive integers, 1, 2, 3, and so forth.

In general, to define a function in R, we use the form

```
function_name =   function (param1, param2, . . .)
    {
    . . .
    }
```

If we want the function to do different things depending on satisfying a certain condition, we write

```
➢ if (condition) statement1 else statement2.
```

Possible conditions include `X>3` and `name=="Mike"`. Note that, in R, the symbol `==` means "is equivalent to," while the symbol `=` means "assign the value of the expression on the right of the equal sign to the variable on the left."

Exercise 2.12.* Write an R function that will compute the number of ways we can choose *m* from *n* things.

2.2.2. Back to the Binomial

We used horses in this last example, but the same reasoning can be applied to coins or survivors in a clinical trial.[4] What is the probability of five heads in ten tosses? What is the probability that five of ten breast cancer patients will still be alive after six months?

We answer these questions in two stages. First, what is the number of different ways we can get five heads in ten tosses? We could have thrown HHHHHTTTTT or HHHHTHTTTT, or some other combination of five heads and five tails for a total of 10 choose 5 or 10!/(5!5!) ways. The probability the first of these events occurs—five heads followed by five tails—is $(\frac{1}{2})^{10}$. Combining these results yields

$$\Pr\{5 \text{ heads in 10 throws of a fair coin}\} = \binom{10}{5}\left(\frac{1}{2}\right)^{10}$$

We can generalize the preceding to an arbitrary probability of success p, $0 \leq p \leq 1$. The probability of failure is $1 - p$. The probability of k successes in n trials is given by the binomial formula

$$\binom{n}{k}(p)^k(1-p)^{n-k} \quad \text{for } 0 \leq k \leq n$$

Exercise 2.13. What is the probability of getting at least one head in six flips of a fair coin?

2.2.3. The Problem Jury

At issue in *Ballew v. Georgia*[5] brought before the Supreme Court in 1978 was whether the all-white jury in Ballew's trial represented a denial of Ballew's rights.[6] In the 1960s and 1970s, United States' courts held uniformly that the use of race, gender, religion, or political affiliation to bar

---

[4] If, that is, the probability of survival is the same for every patient. When there are obvious differences from trial to trial—for example, one subject is an otherwise healthy 35-year-old male, the other an elderly 89-year-old who has just recovered from pneumonia—this simple binomial model would not apply.

[5] 435 U.S. 223, 236–237 (1978).

[6] Strictly speaking, it is not the litigant but the potential juror whose rights might have been interfered with. For more on this issue, see Chapter 2 of *Applications of Statistics in the Courtroom*, by Phillip Good, Chapman and Hall, 2001.

citizens from jury service would not be tolerated. In one case in 1963 in which I assisted the defense on appeal, we were able to show that only one black had served on some 163 consecutive jury panels. In this case, we were objecting—successfully—to the methods used to select the jury. In *Ballew V. Georgia*, the defendant was not objecting to the methods but to the composition of the specific jury that had judged him at trial.

In the district in which Ballew was tried, blacks comprised 10% of the population, but Ballew's jury was entirely white. Justice Blackmun wanted to know what the probability was that a jury of 12 persons selected from such a population in accordance with the law would fail to include members of the minority.

If the population in question is large enough, say, a hundred thousand or so, we can assume that the probability of selecting a nonminority jury-person is a constant 90 out of 100. The probability of selecting two non-minority persons in a row according to the product rule for independent events is 0.9×0.9 or 0.81. Repeating this calculation ten more times, once for each of the remaining ten jurypersons, we get a probability of $0.9 \times 0.9 \times \cdots \times 0.9 = 0.28243$ or 28%.

Not incidentally, Justice Blackmun made exactly this same calculation and concluded that Ballew had not been denied his rights.

2.2.4. Properties of the Binomial

Suppose we send out several hundred individuals to interview our customers and find out if they are satisfied with our products. Each individual has the responsibility of interviewing exactly ten customers. Collating the results, we observe several things:

- Out of every 1000 customers 740 reported they were satisfied.
- Results varied from interviewer to interviewer.
- About 6% of the samples included no dissatisfied customers.
- A little more than 2% of the samples included six or more dissatisfied customers.
- The median number of satisfied customers per sample was seven.
- The modal number of satisfied customers per sample was eight.

When we reported these results to our boss, she only seemed interested in the first of them. "Results always vary from interviewer to interviewer, from sample to sample. And the percentages you reported, apart from the 74% satisfaction rate, are immediate consequences of the binomial distribution."

Clearly, our boss was familiar with the formula for k successes in n trials given in Section 2.2.2. From our initial finding, she knew that $p = 0.74$. Thus,

$$\text{Pr}\{k \text{ satisfied customers in sample of } 10\}$$

$$= \binom{10}{k}(0.74)^k (0.26)^{n-k} \quad for\ 0 \le k \le 10$$

To find the median of this distribution, we can use R as follows:

```
➢ qbinom(.50,10,0.74)
[1] 7
```

To find the proportion of samples with no dissatisfied customers:

```
➢ dbinom(10,10,0.74)
[1] 0.0492399
```

To find the proportion of samples with four or less satisfied customers:

```
➢ pbinom(4,10,0.74)
```

To display this binomial distribution completely and find its mode:

```
➢ # create a vector called binom.vals to hold the
  possible values the results of 10 binomial trials might
  take.
➢ binom.vals = 0:10
➢ binom.prop = dbinom(binom.vals,10,0.74)
➢ #display the results
➢ binom.prop
 [1] 1.411671e-06 4.017833e-05 5.145917e-04 3.905619e-03
     1.945299e-02
 [6] 6.643943e-02 1.575807e-01 2.562851e-01 2.735350e-01
     1.730051e-01
[11] 4.923990e-02

➢ max(binom.prop)
[1] 2.735350e-01
```

Note that the elements of vectors are numbered consecutively beginning at [1]. The mode of the distribution, the outcome with the greatest proportion, is the one located at position 9 in the vector.

```
➢ binom.vals[9]
[1]  8
```

This result tells us that this outcome is 8 successes. The scientific notation used by R in reporting very small or very large values may be difficult to read. Try

```
➢ round(binom.prop,6)
```

for a slightly different view. The 6 asks for six places to the right of the decimal point.

To find the mean or *expected value* of this binomial distribution, let us first note that the computation of the arithmetic mean can be simplified when there are a large number of ties by multiplying each distinct number k in a sample by the frequency f_k with which it occurs; $\overline{X} = \Sigma_k k f_k$. We can have only 11 possible outcomes as a result of our interviews: $0, 1, \ldots$, or 10 satisfied customers. We know from the binomial distribution the frequency f_i with which each outcome may be expected to occur; the population mean is given by the formula

$$\sum_{i=0}^{10} i \binom{10}{i} (p)^i (1-p)^{10-i}$$

For our binomial distribution, we have the values of the variable in the vector binom.vals and their proportions in binom.prop. To find the expected value of our distribution using R, type

```
➢ sum(binom.vals*binom.prop)
[1]  7.4
```

This result, we notice, is equal to 10*0.74 and, more generally, the expected value of the binomial distribution is equal to the product of the sample size and the probability of success at each trial.

Warning: In the preceding example, we assumed that our sample of 1000 customers was large enough that we could use the proportion of successes in that sample, 740 out of 1000, as if it were the true proportion in the entire distribution of customers. Because of the variation inherent in our observations, the true proportion might have been greater or less than our estimate.

Exercise 2.14. Which is more likely—observing two or more successes in eight trials with a probability of one-half of observing a success in each trial, or observing three or more successes in seven trials with a probability of 0.6 of observing a success? Which set of trials has the greater expected number of successes?

Exercise 2.15. Show without using algebra that if X and Y are independent identically distributed binomial variables $B(n, p)$, then $X + Y$ is distributed as $B(2n, p)$.

Unless we have a large number of samples, the observed or *empirical distribution* may differ radically from the expected or theoretical distribution.

Exercise 2.16. One can use the R function rbinom to generate the results of binomial sampling. For example, ➤binom=rbinom(100,10,0.6) will store the results of 100 binomial samples of 10 trials each with probability 0.6 of success in the vector binom. Use R to generate such a vector of sample results and to produce graphs to be used in comparing the resulting empirical distribution with the corresponding theoretical distribution. (See Figure 2.2.)

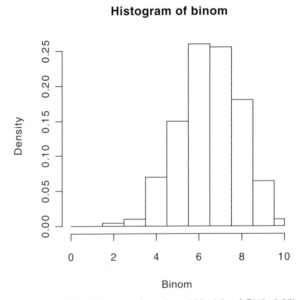

Histogram of binom

Figure 2.2. Histogram, based on 200 trials of $B(10, 0.65)$.

2.2.5. Multinomial

Suppose now that reporters were to take a survey before an election in which multiple candidates were competing for the same office. The reporters would be interested in not only whether or not votes were going to be cast for our candidate (a binomial) but which candidate the votes were going to go to (a *multinomial*). A proportion p_i of the population intends to vote for the ith candidate, where $\Sigma_i p_i = 1$. The reporter is going to use the frequencies $\{f_i\}$ he observes in his survey to estimate the unknown population proportions $\{p_i\}$.[7]

In another application of the multinomial, we might want to do a survey of consumers and have them try washing with our soap. Afterward, we would ask them to state their degree of satisfaction on a five-point scale and, at the same time, state their degree of satisfaction with their present soap. With the comparative data in hand, we could create side-by-side bar charts of the two sets of preferences to use in our advertising.

2.3. CONDITIONAL PROBABILITY

Conditional probability is one of the most difficult of statistical concepts, not so much to understand as to accept in all its implications. Recall that mathematicians arbitrarily assign a probability of 1 to the result that something will happen—the coin will come up heads or tails—and 0 to the probability that nothing will occur. But real life is more restricted: a series of past events has preceded our present and every future outcome is conditioned on this past. Consequently, we need a method whereby the probabilities of just the remaining possibilities sum to 1.

We define the *conditional probability* of an event A given another event B, written $P(A|B)$, to be the ratio $P(A \text{ and } B)/P(B)$. To show how this would work, suppose we are playing craps, a game in which we throw two six-sided die. Clearly, there are $6 \times 6 = 36$ possible outcomes. One (and only one) of these 36 outcomes is snake eyes, a 1 and a 1.

Now, suppose we throw one die at a time (a method that is absolutely forbidden in any real game of craps, whether in the Bellagio or a back alley) and a 1 appears on the first die. The probability that we will now roll snake eyes, that is, that the second die will reveal a 1 also, is 1 out of 6 possibilities or $(1/36)/(1/6) = 6/36 = 1/6$.

The conditional probability of rolling a total of 7 spots on the two dice is 1/6. And the conditional probability of the spots on the two dice summing

---

[7] The choice of letter used for the index is unimportant. $\Sigma_i p_i$ means the same as $\Sigma_k p_k$.

to 11, another winning combination, is 0. Yet before we rolled the two dice, the unconditional probability of rolling snake eyes was 1 out of 36 possibilities and the probability of 11 spots on the two dice was 2/36 (a 5 and a 6 or a 6 and a 5).

Now, suppose I walk into the next room where I have two decks of cards. One is an ordinary deck of 52 cards, half red and half black. The other is a trick deck in which all the spots on the cards are black. I throw a coin—I'm still in the next room so you don't get to see the result of the coin toss—and if the coin comes up heads I stick the trick deck in my pocket, otherwise I take the normal deck. Now, I come back into the room and offer to show you a card chosen at random from the deck in my pocket. The card has black spots. Would you like to bet on whether or not I'm carrying the trick deck?

[STOP: Think about your answer before reading further.]

Common sense would seem to suggest that the odds are still only 50–50 that it's the trick deck I'm carrying. You didn't really learn anything from seeing a card that could have come from either deck. Or did you?

Let's use our conditional probability relation to find out whether the odds have changed. First, what do we know? As the deck was chosen at random, we know that the probability of the card being drawn from the trick deck is the same as the probability of it being drawn from the normal one:

$$P(T^c) = P(T) = \tfrac{1}{2}$$

Here, T denotes the event that I was carrying a trick deck and T^c denotes the complementary event that I was carrying the normal deck.

We also know two conditional probabilities. The probability of drawing a black card from the trick deck is, of course, 1 while that of drawing a black card from a deck that has equal numbers of black and red cards is $\tfrac{1}{2}$. In symbols, $P(B|T) = 1$ and $P(B|T^c) = \tfrac{1}{2}$.

What we'd like to know is whether the two conditional probabilities $P(T|B)$ and $P(T^c|B)$ are equal. We begin by putting the two sets of facts we have together, using our conditional probability relation, $P(B|T) = P(T$ and $B)/P(T)$.

We know two of the values in the first relation, $P(B|T)$ and $P(T)$, and so we can solve for $P(B$ and $T) = P(B|T) P(T) = 1 \times \tfrac{1}{2}$. Similarly, $P(B$ and $T^c) = P(B|T^c) P(T^c) = \tfrac{1}{2} \times \tfrac{1}{2} = \tfrac{1}{4}$.

Take another look at our Venn diagram in Figure 2.1. All the events in outcome B are either in A or in its complement A^c. Similarly, $P(B) = P(B$ and $T) + P(B$ and $T^c) = \tfrac{1}{2} + \tfrac{1}{4} = \tfrac{3}{4}$.

We now know all we need to know to calculate the conditional probability $P(T|B)$ for our conditional probability relation can be rewritten to interchange the roles of the two outcomes, giving $P(T|B) = P(B \text{ and } T)/P(B)$ $= (\frac{1}{2}) / (\frac{3}{4}) = \frac{2}{3}$. By definition $P(T^c|B) = 1 - P(T|B) = \frac{1}{3} < P(T|B)$.

The odds have changed. Before I showed you the card, the probability of my showing you a black card was $1 \times \frac{1}{2} + \frac{1}{2} \times \frac{1}{2}$ or $\frac{3}{4}$. When I showed you a black card, the probability it came from a black deck was $\frac{1}{2}$ divided by $\frac{3}{4}$ or $\frac{2}{3}$!

Exercise 2.17. If R denotes a red card, what would be $P(T|R)$ and $P(T^c|R)$?

A TOO REAL EXAMPLE

Think the previous example was artificial? That it would never happen in real life? My wife and I just came back from a car trip. On our way up the coast, I discovered that my commuter cup leaked, but, desperate for coffee, I wrapped a towel around the cup and persevered. Not in time, my wife noted, pointing to the stains on my jacket.

On our way back down, I lucked out and drew the cup that didn't leak. My wife congratulated me on my good fortune and then, ignoring all she might have learned had she read this text, proceeded to drink from the remaining cup! So much for her new Monterey Bay Aquarium sweat shirt.

2.3.1. Market Basket Analysis

Many supermarkets collect data on purchases using barcode scanners located at the check-out counter. Each transaction record lists all items bought by a customer on a single purchase transaction. Executives want to know whether certain groups of items are consistently purchased together. They use this data for adjusting store layouts (placing items optimally with respect to each other), for cross-selling, for promotions, for catalog design, and to identify customer segments based on buying patterns.

If a supermarket database has 100,000 point-of-sale transactions, out of which 2000 include both items A and B and 800 of these include item C, the *association rule* "if A and B are purchased then C is purchased on the same trip" has a *support* of 800 transactions (alternatively, 0.8% = 800/100,000) and a *confidence* of 40% (= 800/2000).

Exercise 2.18. Suppose you have the results of a market basket analysis in hand. (a) If you wanted an estimate of the probability that a customer will purchase anchovies, would you use the support or the confidence? (b) If you wanted an estimate of the probability that a customer carrying anchovies will also purchase hot dogs, would you use the support or the confidence?

2.3.2. Negative Results

Suppose you were to bet on a six-horse race in which the horses carried varying weights on their saddles. As a result of these handicaps, the probability that a specific horse will win is exactly the same as that of any other horse in the race. What is the probability that your horse will come in first?

Now suppose, to your horror, a horse other than the one you bet on finishes first. No problem; you say, "I bet on my horse to place," that is, you bet it would come in first or second. What is the probability you still can collect on your ticket? That is, what is the conditional probability of your horse coming in second when it did not come in first?

One of the other horses did finish first, which leaves five horses still in the running for second place. Each horse, including the one you bet on, has the same probability to finish second, so the probability you can still collect is one out of five. Agreed?

Just then, you hear the announcer call out that the horses are about to line up for the second race. Again there are six horses and each is equally likely to finish first. What is the probability that if you bet on a horse to place in the second race that you will collect on your bet? Is this $\frac{1}{6} + \frac{1}{5}$?

There are three ways we can arrive at the correct answer when all horses are equally fast:

1. We could notice that the probability that our horse will finish second is exactly the same as the probability that it will finish first (or the probability that it will finish dead last, for that matter). As these are mutually exclusive outcomes, their probabilities may be added. The probability is $\frac{2}{6}$ that your horse finishes first or second.

2. We could list all 6! mutually exclusive outcomes of the race and see how many would lead to our collecting on our bet—but this would be a lot of work.

3. Or, we could trace the paths that lead to the desired result. For example, either our horse comes in first with probability $\frac{1}{6}$ or it does not, with probability $\frac{5}{6}$. If it doesn't come in first, it might still come in second with probability $\frac{1}{5}$. The overall probability of your collecting on your bet is Pr{your horse wins} + Pr{your horse doesn't win} * Pr{your horse is first among the nonwinning horses} $= \frac{1}{6} + \frac{5}{6} * \frac{1}{5} = \frac{2}{6}$.

Exercise 2.19. Suppose ten people are in a class. What is the probability that no two of them were born on the same day of the week? What is the probability that all of them were born in nonoverlapping four-week periods?

Exercise 2.20. A spacecraft depends on five different mission-critical systems. If any of these systems fail, the flight will end in catastrophe. Taken on an individual basis, the probability that a mission-critical system will fail during the flight is $\frac{1}{10}$. (a) What is the probability that the flight will be successful?

NASA decides to build in redundancies. Every mission-critical system has exactly one back-up system that will take over in the event that the primary system fails. The back-up systems have the same probability of failure as the primaries. (b) What is the probability that the flight will be successful?

Exercise 2.21. A woman sued a Las Vegas casino alleging the following. She asked a security guard to hold her slot machine while she hit the buffet; he let somebody else use "her" machine while she was gone; that "somebody else" hit the jackpot; that jackpot was rightfully hers. The casino countered that jackpots were triggered by a random clock keyed to the $\frac{1}{1000}$th of a second; thus, even had the woman been playing the machine continuously, she might not have hit the jackpot. How would you rule if you were a judge?

Exercise 2.22. In the United States in 1985, there were 2.1 million deaths from all causes, compared to 1.7 million in 1960. Does this mean it was safer to live in the United States in the 1960s than in the 1980s?

Exercise 2.23. You are a contestant on "Let's Make a Deal." Monty offers you a choice of three different curtains and tells you there is a brand new automobile behind one of them plus enough money to pay the taxes in case you win the car. You tell Monty you want to look behind curtain number 1. Instead, he throws back curtain number 2 to reveal . . . a child's toy. "Would you like to choose curtain number 3 instead of number 1?" Monty asks. Well, would you?

2.4. INDEPENDENCE

A key element in virtually all the statistical procedures we will consider in this text is that the selection of one member of a sample takes place *independently* of the selection of another. In discussing the game of craps, we assumed that the spots displayed on the first die were *independent* of the spots displayed on the second. When statistics are used, either we:

1. Assume observations are independent.
2. Test for independence.
3. Try to characterize the nature of the dependence (Chapter 7).

Two events or observations are said to be independent of one another, providing knowledge of the outcome or value of the one gives you no information regarding the outcome or value of the other.

In terms of conditional probabilities, two events A and B are independent of one another providing that $P(A|B) = P(A)$; that is, our knowledge that B occurred does not alter the likelihood of A. We can use this relation to show that if A and B are independent, then the probability they will both occur is the product of their separate probabilities, $P(A \text{ and } B) = P(A) * P(B)$. From the definition of conditional probability, $P(A \text{ and } B) = P(A) * P(A \text{ and } B|A) = P(A) * P(B|A) = P(A) * P(B)$.

Warning: Whether events are independent of one another will depend on the context. Imagine that three psychiatrists interview the same individual who we shall suppose is a paranoid schizophrenic. The interviews take place at different times and the psychiatrists are not given the opportunity to confer with each other either before or after the interviews take place.

Suppose now that these psychiatrists are asked for their opinions on (a) the individual's sanity, and, having been informed of the patient's true condition, (b) their views on paranoid schizophrenia. In the first case, their opinions will be independent of one another; in the second case, they will not.

Exercise 2.24. Can two independent events be mutually exclusive?

Exercise 2.25. Draw a Venn diagram depicting two independent events, one of which is twice as likely to occur as the other.

Exercise 2.26. Do the following constitute independent observations?

A. Several students sitting together at a table are asked who their favorite movie actress is.

B. The number of abnormalities in each of several tissue sections taken from the same individual.

C. Opinions of several individuals whose names you obtained by sticking a pin through a phone book, and calling the "pinned" name on each page.

D. Opinions of an ardent Democrat and an ardent Republican.

E. Today's price in Australian dollars of the German mark and the Japanese yen.

Exercise 2.27. Based on the results in the following *contingency tables*, would you say that sex and survival are independent of one another in Table A? in Table B?

Table A		
	Alive	Dead
Men	15	5
Women	15	10

Table B		
	Alive	Dead
Men	15	10
Women	12	8

Exercise 2.28. Provide an example in which an observation X is independent of the value taken by an observation Y, X is independent of a third observation Z, and Y is independent of Z, but X, Y, and Z are dependent.

2.5. APPLICATIONS TO GENETICS

All the information needed to construct an organism, whether a pea plant, a jellyfish, or a person, is encoded in its genes. Each gene contains the information needed to construct a single protein. Each of our cells has two copies of each gene, one obtained from our mother and one from our father. We will contribute just one of these copies to each of our offspring. Whether it is the copy we got from our father or the one from our mother is determined entirely by chance.

You could think of this as flipping a coin, one side says "mother's gene," the other side says "father's gene." Each time a sperm is created in our testis or an ovum in our ovary, the coin is flipped.

There may be many forms of a single gene; each such form is called an allele. Some alleles are defective, incapable of constructing the necessary protein. For example, my mother was rh– meaning that both her copies of the rh gene were incapable of manufacturing the rh protein that is found in red blood cells. This also means that the copy of the rh gene I obtained from my mother was rh–. But my blood tests positive for the rh protein, which means that the rh gene I got from my father was rh+.

Exercise 2.29. The mother of my children was also rh–. What proportion of our children would you expect to be rh–?

Exercise 2.30. Sixteen percent of the population of the United States are rh–. What percentage do you expect to have at least one rh– gene? (Remember, as long as a person has even one rh+ gene, they can manufacture the rh protein.)

The gene responsible for making the A and B blood proteins has three alleles, A, B, and O. A person with two type O alleles will have blood type O. A person with one A allele and one B allele will have blood type AB. Only 4% of the population of the United States have this latter blood type.

Our genes are located on chromosomes. The chromosomes come in pairs, one member of each pair being inherited from the father and one from the mother. Your chromosomes are passed onto the offspring independently of one another.

Exercise 2.31. The ABO and rh genes are located on different chromosomes. What percentage of the population of the United States would you expect to have the AB rh+ blood type?

Exercise 2.32. Forty-five percent of the population of the United States have type O blood. That is, they do not test positive for either the A or the B protein. What percentage of the population do you expect to have at least one O allele?

2.6. SUMMARY AND REVIEW

In this chapter, we introduced the basics of probability theory and independence and considered the properties of a *discrete* probability distribution, the binomial, and applied the elements of probability to genetics. We learned additional R commands for controlling program flow (if, else, while) and for displaying properties of binomial distributions. We also learned how to create our own special purpose functions.

Exercise 2.33. Make a list of all the italicized terms in this chapter. Provide a definition for each one along with an example.

Exercise 2.34. (Read and reread carefully before even attempting an answer.) A magician has three decks of cards, one with only red cards, one that is a normal deck, and one with only black cards. He walks into an adjoining room and returns with only a single deck. He removes the top card from the deck and shows it to you. The card is black. What is the probability that the deck from which the card came consists only of black cards?

Exercise 2.35. An integer number is chosen at random. What is the probability that it is divisible by 2? What is the probability that it is divisible by 17? What is the probability that it is divisible by 2 and 17? What is the probability that it is divisible by 2 or 17? (*Hint*: A Venn diagram would be a big aid in solving this last part.)

Exercise 2.36. Pete, Phil, and Myron are locked in a squash court after hours with only a Twinkie and a coin between them. The only thing all three can agree on is that they want a whole Twinkie or nothing. Myron suggests that Pete and Phil flip the coin, and that the winner flips a coin with him to see who gets the Twinkie. Phil who is a graduate student in statistics says this is unfair. Is it unfair and why? How would you decide who gets the Twinkie?

Exercise 2.37. *People v Collins*[8], concerned a prosecution of an African-American and his Caucasian wife for robbery. The victim testified her purse was snatched by a girl with a blond ponytail; a second witness testified he saw a blond girl, ponytail flying, enter a yellow convertible being driven by an African-American with a beard and mustache. Neither witness could identify the suspects directly. In an attempt to prove the defendants were in fact the persons who had committed the crime, the prosecutor called a college instructor of mathematics to establish that, assuming the robbery were committed by a Caucasian woman with a blond ponytail who left the scene in a yellow Lincoln accompanied by a African-American with a beard and mustache, there was an overwhelming probability the crime was committed by any couple answering to such distinctive characteristics.

Assume you were on the jury. What would you decide and why?

3

DISTRIBUTIONS

In this chapter, you'll learn to recognize and describe the probability distributions of numerical observations made on random selections from a population. You'll learn methods for estimating the parameters of these distributions and for testing hypotheses.

3.1. DISTRIBUTION OF VALUES

Life constantly calls upon us to make decisions. Should penicillin or erythromycin be prescribed for an infection? Which fertilizer should be used to get larger tomatoes? Which style of dress should our company manufacture (i.e., which style will lead to greater sales)?

My wife often asks me what appears to be a very similar question: "Which dress do you think I should wear to the party?" But this question is really quite different from the others as it asks what a specific individual, me, thinks about dresses that are to be worn by another specific individual, my wife. All the other questions reference the behavior of a yet-to-be-determined individual selected *at random* from a population.

Is Alice taller than Peter? It won't take us long to find out: We just need to put the two back to back or measure them separately. The somewhat dif-

Introduction to Statistics Through Resampling Methods and R/S-PLUS®, By Phillip I. Good
Copyright © 2005 by John Wiley & Sons, Inc.

ferent question, "Are girls taller than boys?", is not answered quite so readily. How old are the boys and girls? Generally, but not always, girls are taller than boys until they enter adolescence. Even then, what may be true in general for boys and girls of a specified age group may not be true for a particular girl and a particular boy.

Put in its most general and abstract form, what we are asking is whether a numerical observation X made on an individual drawn at random from one population will be larger than a similar numerical observation Y made on an individual drawn at random from a second population.

3.1.1. Cumulative Distribution Function

Let $F_W[w]$ denote the probability that the numerical value of an observation from the distribution W will be less than or equal to w. $F_W[w]$ is a monotone nondecreasing function. In symbols, if $w < z$, then $0 = F_W[-\infty] \leq F_W[w] = \Pr\{W \leq w\} \leq \Pr\{W \leq z\} = F_W[z] \leq F_W[\infty] = 1$.

Two such *cumulative distribution functions* F_X and G_X are depicted in Figure 3.1.

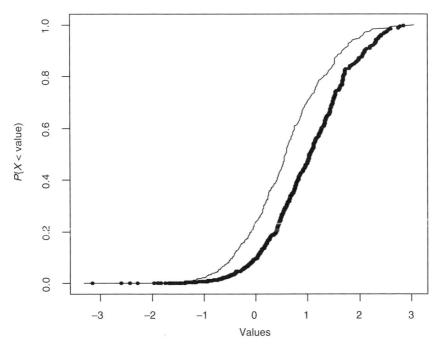

Figure 3.1. Two cumulative distributions that differ by a shift in the median value.

Note the following in this figure:

1. F_X is to the left of G_X. As can be seen by drawing lines perpendicular to the value-axis, $F_X[x] > G_X[x]$ for all values of X. As can be seen by drawing lines perpendicular to the percentile-axis, all the percentiles of the cumulative distribution G are smaller than the percentiles of the cumulative distribution F.

2. Most of the time an observation taken at random from the distribution of X will be smaller if that distribution has cumulative distribution F than if it has cumulative distribution G.

3. Still, there is a nonzero probability that an observation from F_X will be larger than one from G_X.

Many treatments act by shifting the distribution of values, as shown in Figure 3.1. The balance of this chapter and the next is concerned with the detection of such treatment effects. The possibility exists that the actual effects of treatment are more complex than is depicted in Figure 3.1. In such cases (e.g., see Figure 3.2), the introductory methods described in this text may not be immediately applicable.

Exercise 3.1. Is it possible that an observation drawn at random from the distribution F_X depicted in Figure 3.1 could have a value larger than an observation drawn at random from the distribution G_X?[1]

3.1.2 Empirical Distribution Function

Suppose we have collected a sample of n observations $x_1, x_2, \ldots, x_n$. The *empirical cumulative distribution function* $F_n[x]$ is equal to the number of observations that are less than or equal to x divided by the size of our sample, n. If we've sorted the sample so that $x_1 \leq x_2 \leq \ldots \leq x_n$, then

$$F_n[x] = \begin{cases} 0 & \text{if } x < x_1 \\ k/n & \text{if } x_k \leq x < x_{k+1}, \text{ for } 1 \leq k \leq (n-1) \\ 1 & \text{if } x > x_n \end{cases}$$

If these observations all come from the same population distribution **F** and are independent of one another, then as the sample size n gets larger,

---

[1] If the answer to this exercise is not immediately obvious—you'll find the correct answer at the end of this chapter—you should reread Chapters 1 and 2 before proceeding further.

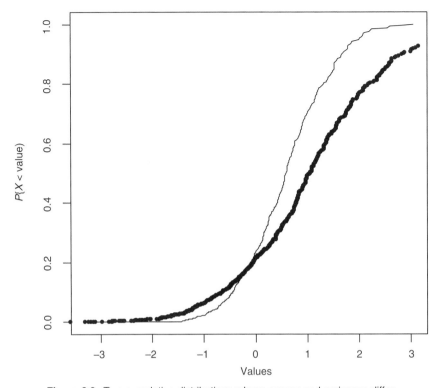

Figure 3.2. Two cumulative distributions whose means and variances differ.

F_n will begin to resemble **F** more and more closely. We illustrate this point in Figure 3.3 with samples of size 10, 100, and 1000 all taken from the same distribution.

Figure 3.3 reveals what you will find in your own samples in practice: The distance (or fit) between the empirical and theoretical distributions is best in the middle of the distribution near the median and worst in the tails.

3.2. DISCRETE DISTRIBUTIONS

We need to distinguish between *discrete* random observations like the binomial and the Poisson made when recording numbers of events and the *continuous* random observations that are made when taking measurements.

Discrete random observations usually take only integer values (positive or negative) with nonzero probability. That is,

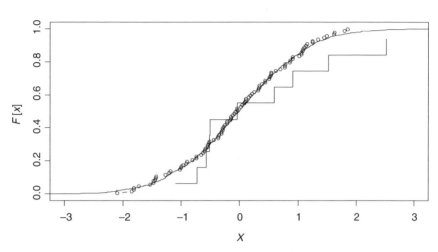

Figure 3.3. Three empirical distributions based on samples of size 10, 100, and 1000 independent observations from the same population.

$$\Pr\{X = k\} = f_k \quad \text{if } k = \{\ldots, -1, 0, 1, 2, \ldots\}$$
$$= 0 \quad \text{otherwise}$$

The cumulative distribution function $F[x]$ is $\Sigma_{k \leq x} f_k$.

Recall from the preceding chapter (Section 2.2.2) that binomial variables had probabilities

$$\Pr\{X = k | n, p\} = \binom{n}{k} (p)^k (1 - p)^{n-k} \quad \text{for } k = \{0, 1, 2, \ldots, n\}$$
$$= 0 \qquad \qquad \text{otherwise}$$

where n denoted the number of independent trials and p was the probability of success in each trial. The cumulative distribution function of a binomial is a step function, equal to zero for all values less than 0 and equal to one for all values greater than or equal to n.

Though it seems obvious that the mean of a sufficiently large number of sets of n binomial trials each with a probability p of success will be equal to np, many things that seem obvious in mathematics aren't. Besides, if it's that obvious, we should be able to prove it.

If a variable X takes a discrete set of values $\{\ldots, 0, 1, \ldots, k, \ldots\}$ with corresponding probabilities $\{\ldots, f_0, f_1, \ldots, f_k, \ldots\}$, its mean or *expected value*, written EX, is given by the summation $\ldots + 0f_0 + 1f_1 + \ldots + kf_k + \ldots$, which we may also write as $\Sigma_k k f_k$. For a binomial variable, this sum is

$$EX = \sum_{k=0}^{n} k \binom{n}{k} (p)^k (1-p)^{n-k}$$

$$= \sum_{k=0}^{n} k \frac{n(n-1)!}{k!(n-k)!} p(p)^{k-1} (1-p)^{n-1-(k-1)}$$

Note that the term on the right is equal to zero when $k = 0$, so we can start the summation at $k = 1$. Factoring n and p outside the summation and using the k in the numerator to reduce $k!$ to $(k-1)!$, we have

$$EX = np \sum_{k=1}^{n} \frac{(n-1)!}{(k-1)![(n-1)-(k-1)]!} (p)^{k-1} (1-p)^{(n-1)-(k-1)}$$

If we change the notation a bit, letting $j = k - 1$ and $m = n - 1$, this can be expressed as

$$EX = np \sum_{j=0}^{m} \binom{m}{j} (p)^j (1-p)^{m-j}$$

The summation on the right side of the equal sign is of the probabilities associated with *all* possible outcomes for the binomial random variable $B(n, p)$, so it must be equal to 1. Thus, $EX = np$, a result that agrees with our intuitive feeling that, in the long run, the number of successes should be proportional to the probability of success.

Suppose $EX = \theta$ (pronounced theta). The *variance* of X is defined as $\mathrm{Var}(X) = E(X - \theta)^2$. In contrast to the *sample variance* defined in Chapter 1, $\mathrm{Var}(X)$ stands for a purely hypothetical value. For observations from a discrete distribution, $\mathrm{Var}(X) = \Sigma_k (k - \theta)^2 f_k$.

Exercise 3.2. (for math and statistics majors and the intensely curious only) Show that the variance of a binomial variable is $np(1 - p)$.

Exercise 3.3. Is the binomial distribution *symmetric* about its mean? Do its mean and median coincide? Does it have more than one mode? [*Hint:* Use the program provided in Section 2.2.4 to display the binomial distribution for various probabilities of success and numbers of trials.]

Exercise 3.4. Recently, we interviewed ten people and found the majority favored our candidate. Should we conclude that our candidate is sure to win a majority? Support your opinion with numerical values.

Exercise 3.5. Create side-by-side plots of the cumulative distribution functions of the binomial random variables $B(20, 0.5)$ and $B(20, 0.7)$.

We used the following R instructions to create the side-by-side plots you see in Figure 3.3:

```
> plot(sort(y), ppoints(y), type="s", xlab="X", ylab="F[x]")
> points(sort(w), ppoints(w), pch = 21)
> points(sort(z), ppoints(z), type = "l")
```

3.3. POISSON: EVENTS RARE IN TIME AND SPACE

The decay of a radioactive element, an appointment to the United States Supreme Court, and a cavalry officer trampled by his horse have in common that they are relatively rare but inevitable events. They are inevitable, that is, if there are enough atoms, enough seconds or years in the observation period, and enough horses and momentarily careless men. Their frequency of occurrence has a Poisson distribution.

The number of events in a given interval has the Poisson distribution if (a) it is the cumulative result of a large number of independent opportunities each of which has only a small chance of occurring, and (b) events in nonoverlapping intervals are independent.

The intervals can be in space or time. For example, if we seed a small number of cells into a petri dish that is divided into a large number of squares, the distribution of cells per square follows the Poisson. The same appears to be true in the way a large number of masses in the form of galaxies are distributed across a very large universe.

Like the binomial variable, a Poisson variable takes only non-negative integer values. If the number of events X has a Poisson distribution such that we may expect an average of λ events per unit interval, then $\Pr\{X = k\} = \lambda^k e^{-\lambda}/k!$ for $k = 0, 1, 2, \ldots$. For the purpose of testing hypotheses concerning λ as discussed in Chapter 4, we needn't keep track of the times or locations at which the various events occur; the number of events k is *sufficient*.

Exercise 3.6. Show without using algebra that the sum of a Poisson with expected value λ_1 and a second independent Poisson with expected value λ_2 is also a Poisson with expected value $\lambda_1 + \lambda_2$.

3.3.1. Applying the Poisson

John Ross of the Wistar Institute held there were two approaches to biology: the analog and the digital. The analog was served by the scintillation counter: One ground up millions of cells then measured whatever radioactivity was left behind in the stew after centrifugation. The digital was to be found in cloning experiments where any necessary measurements would be done on a cell-by-cell basis.

John was a cloner and, later, as his student, so was I. We'd start out with 10 million or more cells in a 10-milliliter flask and try to dilute them down to one cell per milliliter. We were usually successful in cutting down the numbers to 10 thousand or so. Then came the hard part. We'd dilute the cells down a second time by a factor of 1:100 and hope we'd end up with 100 cells in the flask. Sometimes we did. Ninety percent of the time, we'd end up with between 90 and 110 cells, just as the binomial distribution predicted. But just because you cut a mixture in half (or a dozen, or 100 parts) doesn't mean you're going to get equal numbers in each part. It means the probability of getting a particular cell is the same for all the parts. With large numbers of cells, things seem to even out. With small numbers, chance seems to predominate.

Things got worse when I went to seed the cells into culture dishes. These dishes, made of plastic, had a rectangular grid cut into their bottoms, so they were divided into approximately 100 equal size squares. Dropping 100 cells into the dish meant an average of 1 cell per square. Unfortunately for cloning purposes, this average didn't mean much. Sometimes, 40% or more of the squares would contain two or more cells. It didn't take long to figure out why. Planted at random, the cells obey the Poisson distribution in space. An average of one cell per square means

$$\Pr\{\text{no cells in a square}\} = 1 * e^{-1}/1 = 0.32$$

$$\Pr\{\text{exactly one cell in a square}\} = 1 * e^{-1}/1 = 0.32$$

$$\Pr\{\text{two or more cells in a square}\} = 1 - 0.32 - 0.32 = 0.36$$

Two cells were one too many. A clone or colony must begin with a single cell. I had to dilute the mixture a third time to ensure the percentage of squares that included two or more cells was vanishingly small. Alas, the vast majority of squares were now empty; I was forced to spend hundreds of additional hours peering through the microscope looking for the few squares that did include a clone.

3.3.2. Comparing Empirical and Theoretical Poisson Distributions

R has functions for use with the Poisson similar to those we used in studying the binomial including dpois, ppois, qpois, and rpois.

Exercise 3.7. Generate the results of 100 samples from a Poisson distribution with an expected number of 2 events per interval. Compare the graph of the resulting empirical distribution with that of the corresponding theoretical distribution. Determine the 10th, 50th, and 90th percentiles of the theoretical Poisson distribution.

Exercise 3.8. Show that if $\Pr\{X = k\} = \lambda^k e^{-\lambda}/k!$ for $k = 0, 1, 2, \ldots$; that is, if X is a Poisson variable, then the expected value of $X = \Sigma_k k \Pr\{X = k\} = \lambda$.

Exercise 3.9. In subsequent chapters, we will learn how the statistical analysis of trials of a new vaccine is often simplified by assuming that the number of infected individuals follows a Poisson rather than a binomial distribution. To see how accurate an approximation this might be, compare the cumulative distribution functions of a binomial variable, $B(100, 0.01)$ and a Poisson variable, $P(1)$, over the range 0 to 100.

3.4. CONTINUOUS DISTRIBUTIONS

The vast majority of the observations we make are on a continuous scale even if, in practice, we only can make them in discrete increments. For example, a man's height might actually be 1.835421117 meters, but we are not likely to record a value with that degree of accuracy (nor want to). If one's measuring stick is accurate to the nearest millimeter, then the probability that an individual selected at random will be exactly 2 meters tall is really the probability that his or her height will lie between 1.9995 and 2.0004 meters. In such a situation, it is more convenient to replace the sum of an arbitrarily small number of quite small probabilities with an integral

$$\int_{1.9995}^{2.0004} dF[x] = \int_{1.9995}^{2.0004} f[x]\,dx$$

where $F[x]$ is the cumulative distribution function of the continuous variable representing height and $f[x]$ is its probability density. Note that $F[x]$ is now defined as $\int_{-\infty}^{x} f[y]\,dy$.[2] As with discrete variables, the cumulative distribution function is monotone nondecreasing from 0 to 1, the distinction being that it is a smooth curve rather than a step function.

---

[2] If it's been a long while or never since you had calculus, note that the differential dx or dy is a meaningless index, so any letter will do, just as $\Sigma_k f_k$ means exactly the same thing as $\Sigma_j f_j$.

The mathematical expectation of a continuous variable is $\int_{-\infty}^{\infty} yf[y]\, dy$ and its variance is $\int_{-\infty}^{\infty} (y - EY)^2 f[y]\, dy$.

3.4.1. The Exponential Distribution

The simplest way to obtain continuously distributed random observations is via the same process that gave rise to the Poisson. Recall that a Poisson process is such that events in nonoverlapping intervals are independent and identically distributed. The times[3] between Poisson events follow an exponential distribution:

$$F[t|\lambda] = \Pr\{T \le t \mid \lambda\} = 1 - \exp[-\lambda t] \quad \text{for } t \ge 0, \lambda > 0$$

When t is zero, $\exp[-\lambda t]$ is 1 and $F[t \mid \lambda]$ is 0. As t increases, $\exp[-\lambda t]$ decreases rapidly toward zero and $F[t \mid \lambda]$ increases rapidly to 1. The rate of increase is proportional to the magnitude of the parameter λ. In fact, $\log(1 - F[t \mid \lambda]) = -\lambda t$. The next exercise allows you to demonstrate this for yourself.

Exercise 3.10. Draw the cumulative distribution function of an exponentially distributed observation with parameter λ. Is the median the same as the mean?

Exercise 3.11. (requires calculus) What is the expected value of an exponentially distributed observation with parameter λ?

The times between counts on a Geiger counter follow an exponential distribution. So do the times between failures of manufactured items like light bulbs that rely on a single crucial component.

Exercise 3.12. When you walk into a room, you discover the light in a lamp is burning. Assuming the life of its bulb is exponentially distributed with an expectation of one year, how long do you expect it to be before the bulb burns out? [Many people find they get two contradictory answers to this question. If you are one of them, see Feller, 1966, pp. 11–12.]

Most real-life systems (including that complex system known as a human being) have built-in redundancies. Failure can only occur after a series of *n* breakdowns. If these breakdowns are independent and exponentially dis-

---

[3] Time is almost but not quite continuous. Modern cosmologists now believe that both time and space are discrete with time determined only to the nearest 10^{-23} second.

tributed, all with the same parameter λ, the probability of failure of the total system at time $t > 0$ is

$$f(t) = \lambda \exp(-\lambda t)(\lambda t)^n/n!$$

3.4.2. The Normal Distribution

Figure 1.7 depicts the bell-shaped symmetric frequency curve of a normally distributed population. Its probability density $f(x)$ may be written as

$$f[x|\theta, \sigma^2] = \frac{1}{\sqrt{2\pi\sigma^2}} \exp\left(-\frac{1}{2}\right)\left(\frac{y-\theta}{\sigma}\right)^2 \qquad (3.1)$$

In contrast to the exponential distribution, the normal distribution depends on two parameters: its expected value θ (theta) and its variance σ^2 (sigma-squared).

Exercise 3.13. How do changes in the values of these parameters affect the shape of the normal distribution? [*Hint:* Let $w = (y - \theta)/\sigma$, so the probability density function can be written as $f[x|\theta, \sigma^2] = 1/\sqrt{2\pi\sigma^2} \exp(-\frac{1}{2}w^2).$]

Exercise 3.14. (requires calculus) Show that the expected value of a normal distribution whose density is given by Equation 3.1 is θ and its variance is σ^2.

STATISTICS AND PARAMETERS

A *statistic* is any single value such as the sample mean $(1/n)\Sigma_{k=1}^{n}X_k$ that summarizes some aspect of a sample. A parameter is any single value such as the mean θ of a normal distribution that summarizes some aspect of an entire population. Examples of sample statistics include measures of location and central tendency such as the sample mode, sample median, and sample mean, extrema such as the sample minimum and maximum, and measures of variation and dispersion such as the sample standard deviation. These same measures are considered *parameters* when they refer to an entire population, for example, population mean θ, population range, and population variance σ^2.

In subsequent chapters, we will use sample statistics to estimate the values of population parameters and to test hypotheses about them.

To see why the normal distribution plays such an important role in statistics, please complete Exercise 3.14, which requires you to compute the distribution of the mean of a number of binomial observations. As you increase the number of observations used to compute the mean from 5 to 12 so that each individual observation makes a smaller relative contribution to the total, you will see that the distribution of means looks less and less like the binomial distribution from which they were taken and more and more like a normal distribution. This result will prove to be true regardless of the distribution from which the observations used to compute the mean are taken, providing that this distribution has a finite mean and variance.

Exercise 3.15. Generate five binomial observations based on ten trials with probability of success $p = 0.35$ per trial. Compute the mean value of these five. Repeat this procedure 512 times, computing the mean value each time. Plot the histogram of these means. Compare with the histograms of (a) a sample of 512 normally distributed observations with expected value 3.5 and variance 2.3, (b) a sample of 512 binomial observations each consisting of ten trials with probability of success $p = 0.4$ per trial. Repeat the entire exercise, computing the mean of 12 rather than 5 binomial observations.

Here is the R code you'll need to do this exercise:

```
➢ #Program computes the means of N samples of k
  binomial(10,0.35) variables
➢ #Set number of times to compute a mean
➢ N = 512
➢ #Set number of observations in sample
➢ k = 5
➢ #Create a vector in which to store the means
➢ stat = numeric(N)
➢ #Set up a loop to generate the N means
➢ for (i in 1:N) stat[i] = mean(rbinom (k,10, 0.35))
➢ hist(stat,14)
➢
➢ #Generate a sample of N normally distributed
  observations
➢ sampN = rnorm (N, 3.5, 2.3)
➢ hist (sampN,14)
➢
➢ #Generate a sample of N binomial observations
➢ sampB = rbinom (N, 10, 0.35)
➢ hist (sampB,14)
```

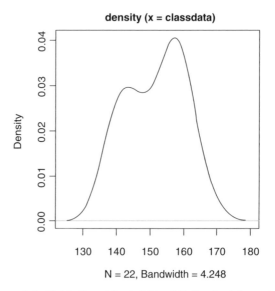

Figure 3.4. Distribution of the heights of California sixth graders.

3.4.3. Mixtures of Normal Distributions

Many real-life distributions strongly resemble mixtures of normal distributions. Figure 3.4 depicts just such an example in the heights of California sixth graders. Though the heights of boys and girls overlap, it is easy to see that the population of sixth graders is composed of a mixture of the two sexes.

3.5. PROPERTIES OF INDEPENDENT OBSERVATIONS

As virtually all the statistical procedures in this text require that our observations be independent of one another, a study of the properties of independent observations will prove rewarding. Recall from the previous chapter that if X and Y are independent observations, then

$$\Pr\{X = k \text{ and } Y = j\} = \Pr\{X = k\}\Pr\{Y = j\}$$

and

$$\Pr\{X = k \mid Y = j\} = \Pr\{X = k\}$$

If X and Y are independent discrete random variables and their expectations exist and are finite,[4] then $E\{X + Y\} = EX + EY$. To see this, suppose that $Y = y$. The conditional expectation of $X + Y$ given $Y = y$ is

$$E\{X + Y \mid Y = y\} = \Sigma_k(k + y)\Pr\{X + Y = k + y \mid Y = y\}$$
$$= \Sigma_k(k + y)\Pr\{X = k\}$$
$$= EX + y\Sigma_k\Pr\{X = k\} = EX + y$$

Taking the average of this conditional expectation over all possible values of Y yields

$$E\{X + Y\} = \Sigma_j(EX + j)\Pr\{Y = j\} = EX * 1 + EY$$

A similar result holds if X and Y have continuous distributions providing that their individual expectations exist and are finite.

Exercise 3.16. Show that for any variable X with a finite expectation, $E(aX) = aEX$, where a is a constant.

Exercise 3.17. Show that the expectation of the mean of n independent identically distributed random variables with finite expectation θ is also θ.

THE CAUCHY DISTRIBUTION: AN EXCEPTION TO THE RULE

It seems obvious that every distribution should have a mean and a variance. But telling mathematicians something is "obvious" only encourages them to find an exception. The Cauchy distribution is just one example: The expression $f(x) = 1/\pi(1 + x)$ for $-\infty < x < \infty$ is a probability density since $\int f(x)\, dx = 1$, but neither its mean nor its variance exists. Does the Cauchy distribution arise in practice? It might if you study the ratio X/Y of two independent random variables each distributed as $N(0, 1)$.

The variance of the sum of two independent variables X and Y is the sum of their variances, providing each of these variances exists and is finite.

---

[4] In real life, expectations almost always exist and are finite—the expectations of ratios are a notable exception.

Exercise 3.18. Given the preceding result, show that the variance of the mean of n independent identically distributed observations is $(1/n)$th of the variance of just one of them. Does this mean that the arithmetic mean is more precise than any individual observation? Does this mean that the sample mean will be closer to the mean of the population from which it is drawn than any individual observation would be, that is, that it would be more accurate?

3.6. TESTING A HYPOTHESIS

Suppose we were to pot a half-dozen tomato plants in ordinary soil and a second half-dozen plants in soil enriched with fertilizer. If we wait a few months, we can determine whether the addition of fertilizer increases the resulting yield of tomatoes, at least as far as these dozen plants are concerned. But can we extend our findings to all tomatoes?

To ensure that we can extend our findings we need to proceed as follows: First, the 12 tomato plants used in our study have to be a *random sample* from a nursery. If we choose only plants with especially green leaves for our sample, then our results can be extended only to plants with especially green leaves. Second, we have to divide the 12 plants into two treatment groups at random. If we subdivide by any other method, such as tall plants in one group and short plants in another, then the experiment would not be about fertilizer but about our choices.

I performed just such an experiment a decade or so ago; only I was interested in the effects vitamin E might have on the aging of human cells in culture. After several months of abject failure—contaminated cultures, spilled containers—I succeeded in cloning human diploid fibroblasts in eight culture dishes. Four of these dishes were filled with a conventional nutrient solution and four held an experimental "life-extending" solution to which vitamin E had been added. All the cells in the dishes came from the same culture so that the initial distribution of cells was completely random.

I waited three weeks with my fingers crossed—there is always a risk of contamination with cell cultures—but at the end of this test period three dishes of each type had survived. I transplanted the cells, let them grow for 24 hours in contact with a radioactive label, and then fixed and stained them before covering them with a photographic emulsion.

Ten days passed and we were ready to examine the autoradiographs: "121, 118, 110, 34, 12, 22." I read and reread these six numbers over and over again. The larger numbers were indicative of more cell generations and an extended lifespan. If the first three generation-counts were from treated colonies and the last three were from untreated, then I had found the fountain of youth. Otherwise, I really had nothing to report.

3.6.1. Analyzing the Experiment

How had I reached this conclusion? Let's take a second, more searching look. First, we identify the primary hypothesis and the alternative hypothesis of interest.

I wanted to assess the life-extending properties of a new experimental treatment with vitamin E. To do this, I had divided my cell cultures into two groups: one grown in a standard medium and one grown in a medium containing vitamin E. At the conclusion of the experiment and after the elimination of several contaminated cultures, both groups consisted of three independently treated dishes.

My primary hypothesis was a *null hypothesis*, that the growth potential of a culture would not be affected by the presence of vitamin E in the media: All the cultures would have equal growth potential. The *alternative* hypothesis of interest was that cells grown in the presence of vitamin E would be capable of many more cell divisions.

Under the null hypothesis, the labels "treated" and "untreated" provide no information about the outcomes: The observations are expected to have more or less the same values in each of the two experimental groups. If they do differ it should only be as a result of some uncontrollable random fluctuation. Thus, if this null or no-difference hypothesis were true, I was free to exchange the labels.

The alternative is a distributional shift like that depicted in Figure 3.1 in which greater numbers of cell generations are to be expected as the result of treatment with vitamin E (though the occasional smaller value cannot be ruled out completely).

The next step is to choose a test statistic that discriminates between the hypothesis and the alternative. The statistic I chose was the sum of the counts in the group treated with vitamin E. *If* the alternative hypothesis is true, most of the time this sum ought to be larger than the sum of the counts in the untreated group. *If* the null hypothesis is true, that is, if it doesn't make any difference which treatment the cells receive, then the sums of the two groups of observations should be approximately the same. One sum might be smaller or larger than the other by chance, but most of the time the two shouldn't be all that different.

The third step is to compute the test statistic for each of the possible re-labelings and compare these values with the value of the test statistic as the data was labeled originally. As it happened, the first three observations—121, 118, and 110—were those belonging to the cultures that received vitamin E. The value of the test statistic for the observations as originally labeled is 349 = 121 + 118 + 110.

I began to rearrange (permute) the observations, randomly reassigning the six labels, three "treated" and three "untreated," to the six observations. For example: treated, 121   118   34, and untreated, 110   12   22. In this particular rearrangement, the sum of the observations in the first (treated) group is 273. I repeated this step until all

$$\binom{6}{3} = 20$$

distinct rearrangements had been examined.[5]

	First Group			Second Group			First Group
				Sum of			
1.	121	118	110	34	22	12	349
2.	121	118	34	110	22	12	273
3.	121	110	34	118	22	12	265
4.	118	110	34	121	22	12	262
5.	121	118	22	110	34	12	261
6.	121	110	22	118	34	12	253
7.	121	118	12	110	34	22	251
8.	118	110	22	121	34	12	250
9.	121	110	12	118	34	22	243
10.	118	110	12	121	34	22	240
11.	121	34	22	118	110	12	177
12.	118	34	22	121	110	12	174
13.	121	34	12	118	110	22	167
14.	110	34	22	121	118	12	166
15.	118	34	12	121	110	22	164
16.	110	34	12	121	118	22	156
17.	121	22	12	118	110	34	155
18.	118	22	12	121	110	34	152
19.	110	22	12	121	118	34	144
20.	34	22	12	121	118	110	68

[5] Determination of the number of relabelings, "6 choose 3" in the present case, is considered in Section 2.2.1.

The sum of the observations in the original vitamin E treated group, 349, is equaled only once and never exceeded in the 20 distinct random re-labelings. If chance alone is operating, then such an extreme value is a rare, only 1-time-in-20, event. If I reject the null hypothesis and embrace the alternative that the treatment is effective and responsible for the observed difference, I only risk making an error and rejecting a true hypothesis once in every 20 times.

In this instance, I did make just such an error. I was never able to replicate the observed life-promoting properties of vitamin E in other repetitions of this experiment. Good statistical methods can reduce and contain the probability of making a bad decision, but they cannot eliminate the possibility.

Exercise 3.19. How was the analysis of the cell culture experiment affected by the loss of two of the cultures due to contamination? Suppose these cultures had escaped contamination and given rise to the observations 90 and 95. What would be the results of a permutation analysis applied to the new, enlarged data set consisting of the following cell counts

Treated 121 118 110 90   Untreated 95 34 22 12?

[*Hint:* To determine how probable an outcome like this is by chance alone, first determine how many possible rearrangements there are. Then list all the rearrangements that are as or more extreme than this one.]

3.6.2. Two Types of Errors

In the preceding example, I risked rejecting the null hypothesis in error 5% of the time. Statisticians call this making a *Type I error* and they call the 5%, the *significance level*. In fact, I did make such an error, as in future experiments, vitamin E proved to be valueless in extending the life span of human cells in culture.

On the other hand, suppose the null hypothesis had been false, that treatment with vitamin E really did extend life span, and I had failed to reject the null hypothesis. Statisticians call this making a *Type II error.*

The consequences of each type of error are quite different and depend on the context of the investigation. Consider Table 3.1 listing the possibilities arising from an investigation of the possible carcinogenicity of a new headache cure.

We may luck out in that our samples support the correct hypothesis, but we always run the risk of making either a Type I or a Type II error. We can't avoid it. If we use a smaller significance level, say, 1%, then if the null hypothesis is false, we are more likely to make a Type II error. If we always

Table 3.1. Decision making under uncertainty

The Facts	Investigator's Decision	
Not a carcinogen	Not a carcinogen	Compound a carcinogen *Type I error*: Manufacturer misses opportunity for profit. Public denied access to effective treatment.
Carcinogen	*Type II error*: Manufacturer sued. Patients die; families suffer.	

accept the null hypothesis, a significance level of 0%, then we guarantee making a Type II error if the null hypothesis is false. This seems kind of stupid: Why bother to gather data if you're not going to use it? But if you read or live Dilbert, then you know this happens all the time.

Exercise 3.20. The nurses have petitioned the CEO of a hospital to allow them to work 12-hour shifts. He wants to please them but is afraid that the frequency of errors may increase as a result of the longer shifts. He decides to conduct a study and to test the null hypothesis that there is no increase in error rate as a result of working longer shifts against the alternative that the frequency of errors increases by at least 30%. Describe the losses associated with Type I and Type II errors.

Exercise 3.21. Design a study. Describe a primary hypothesis of your own along with one or more likely alternatives. The truth or falsity of your chosen hypothesis should have measurable monetary consequences. If you were to test your hypothesis, what would be the consequences of making a Type I error? A Type II error?

Exercise 3.22. Suppose I'm (almost) confident that my candidate will get 60% or more of the votes in the next primary. The alternative that scares me is that she will get 40% or less. To test my confident hypothesis, I decide to interview 20 people selected at random in a shopping mall and reject my hypothesis if 7 or fewer say they will vote for her. What is the probability of my making a Type I error? What is the probability of my retaining confidence in my candidate if only 40% of the general population favor her (i.e., committing a Type II error)? How can I reduce the probability of making a Type II error while keeping the probability of making a Type I error the same?

Exercise 3.23. Individuals were asked to complete an extensive questionnaire concerning their political views and eating preferences. Analyzing the results, a sociologist performed 20 different tests of hypotheses. Unknown to the sociologist, the null hypothesis was true in all 20 cases. What is the probability that the sociologist rejected at least one of the hypotheses at the 5% significance level?

3.7. ESTIMATING EFFECT SIZE

In the previous example, we developed a test of the null hypothesis of no treatment effect against the alternative hypothesis that a positive effect existed. But in many situations, we would also want to know the magnitude of the effect. Does vitamin E extend cell life span by 3 cell generations? By 10? By 15?

In Section 1.6.2, we used the bootstrap to obtain estimates of the precision of an estimate of the sample mean or median (or, indeed, any sample percentile). In Section 1.7.2 we showed how to use the bootstrap estimate of the precision of the sample mean or median (or, indeed, almost any sample statistic) as an estimator of a population parameter. As a by-product, we obtain an *interval estimate* of the corresponding population parameter.

For example, if P_{05} and P_{95} are the 5th and 95th percentiles of the bootstrap distribution of the median of the law school LSAT data you used for Exercise 1.16, then the set of values between P_{05} and P_{95} provides a 90% *confidence interval* for the median of the population from which the data was taken.

Exercise 3.24. Obtain an 85% confidence interval for the median of the population from which the LSAT data was taken.

Exercise 3.25. Can this same bootstrap technique be used to obtain a confidence interval for the 90th percentile of the population? For the maximum value in the population?

3.7.1. Additional Applications

Suppose we have independently collected two random samples and want to know the following:

  Do the populations from which they are drawn have the same means?

If the means are not the same, then what is the difference between them?

To find out, we would let each sample stand in place of the population from which it is drawn, take a series of bootstrap samples separately from each sample, and compute the difference in means each time.

Suppose our data is stored in two vectors called "control" and "treated." The R code to obtain the desired interval estimate would be as follows:

```
➢ #This program selects 400 bootstrap samples from your
   data
➢ #and then produces an interval estimate of the
   difference in population means
➢ #Record group sizes
➢ n = length(control)
➢ m = length(treated)
➢ #set number of bootstrap samples
➢ N = 400
➢ stat = numeric(N) #create a vector in which to store the
   results
➢ #The elements of the vector will be numbered from 1 to
   N
➢ #Set up a loop to generate a series of bootstrap
   samples
➢ for (i in 1:N){
➢ #bootstrap sample counterparts are denoted with a "B"
➢ controlB = sample (control, n, replace=T)
➢ treatB = sample (treated, m, replace=T)
➢ stat[i] = mean(treatB) - mean(controlB)
➢ }
➢ quantile (x=stat, probs = c(.25,.75))
```

With S + Resample instead you can do:

```
➢ boot = bootstrap2(control, data2 = treated, mean, B =
   400)
➢ plot(boot)
➢ quantile(boot$replicates)
```

Yet another example of the bootstrap's application lies in the measurement of the *correlation* or degree of agreement between two variables. The Pearson correlation of two variables X and Y is defined as the ratio of the covariance between X and Y and the product of the standard deviations of

X and Y. The covariance of X and Y is given by the formula $\Sigma_{k=1}^{n}(X_k - \overline{X}.)(Y_k - \overline{Y}.)/(n-1)$.

Recall that if X and Y are independent, then $E(XY) = (EX)(EY)$, so that the expected value of the covariance and hence the correlation of X and Y is zero. If X and Y increase more or less together as do, for example, the height and weight of individuals, their covariance and their correlation will be positive so that we say that height and weight are positively correlated. I had a boss, more than once, who believed that the more abuse and criticism he heaped on an individual the more work he could get out of that individual. Not. Abuse and productivity are negatively correlated; heap on the abuse and work output declines.

The reason we divide by the product of the standard deviations in assessing the degree of agreement between two variables is that it renders the correlation coefficient free of the units of measurement.

If $X = -Y$, so that the two variables are totally dependent, the correlation coefficient, usually represented in symbols by the Greek letter ρ (rho) will be -1. In all cases, $-1 \le \rho \le 1$.

Exercise 3.26. Using the LSAT data from Exercise 1.16 and the bootstrap, obtain an interval estimate for the correlation between the LSAT score and the student's subsequent GPA.

Exercise 3.27. Trying to decide whether to take a trip to Paris or Tokyo, a student kept track of how many euros and yen her dollars would buy. Month by month she found that the values of both currencies were rising. Does this mean that improvements in the European economy are reflected by improvements in the Japanese economy?

3.7.2. Using Confidence Intervals to Test Hypotheses

Suppose we have derived a 90% confidence interval for some parameter, for example, a confidence interval for the difference in means between two populations, one of which was treated and one that was not. We can use this interval to test the hypothesis that the difference in means is 4 units, by accepting this hypothesis if 4 is included in the confidence interval and rejecting it otherwise. If our alternative hypothesis is nondirectional and two-sided, $\theta_A \ne \theta_B$, the test will have a Type I error of $100\% - 90\% = 10\%$.

Clearly, hypothesis tests and confidence intervals are intimately related.

Suppose we test a series of hypotheses concerning a parameter θ. For example, in the vitamin E experiment, we could test the hypothesis that vitamin E has no effect, $\theta = 0$, or that vitamin E increases life span by 25 generations, $\theta = 25$, or that it increases it by 50 generations, $\theta = 50$. In each case, whenever we accept the hypothesis, the corresponding value of the parameter should be included in the confidence interval.

In this example, we are really performing a series of one-sided tests. Our hypotheses are that $\theta = 0$ against the one-sided alternative that $\theta > 0$, that $\theta \leq 25$ against the alternative that $\theta > 25$, and so forth. Our corresponding confidence interval will be one-sided also; we will conclude $\theta < \theta_U$ if we accept the hypothesis $\theta = \theta_0$ for all values of $\theta_0 < \theta_U$ and reject it for all values of $\theta_0 \geq \theta_U$.

Suppose, instead, we wanted to make a comparison between two treatments of unknown value or between two proposed direct-mail appeals for funds. Our null hypothesis would be that it didn't make any difference which treatment or which marketing letter was used; the two would yield equivalent results. Our *two-sided alternative* would be that one of the two would prove superior to the other. Such a two-sided alternative would lead to a two-sided test in which we would reject our null hypothesis if we observed either very large or very small values of the test statistic. Not surprisingly, one-sided tests lead to one-sided confidence intervals and two-sided tests to two-sided confidence intervals.

Exercise 3.28. What is the relationship between the significance level of a test and the confidence level of the corresponding interval estimate?

Exercise 3.29. In each of the following instances would you use a one-sided or a two-sided test?

(a) Determine whether men or women do better on math tests.
(b) Test the hypothesis that women can do as well as men on math tests.
(c) In *Commonwealth v. Rizzo et al.*, 466 F. Supp 1219 (E.D. Pa 1979), help the judge decide whether certain races were discriminated against by the Philadelphia Fire Department by means of an unfair test.

Exercise 3.30. Use the data of Exercise 3.19 to derive an 80% upper confidence bound for the effect of vitamin E to the nearest five cell generations.

3.8. SUMMARY AND REVIEW

In this chapter, we considered the form of four common distributions, two discrete (the binomial and the Poisson) and two continuous (the normal and the exponential). We provided the R functions necessary to generate random samples from the various distributions and to display plots side-by-side on the same graph.

We noted that as sample size increases, the observed or empirical distribution of values more closely resembles the theoretical. The distribution of sample statistics such as the sample mean and sample variance is different from the distribution of individual values. In particular, under very general conditions with moderate-size samples, the distribution of the sample mean will take on the form of a normal distribution. We considered two non-parametric methods—the bootstrap and the permutation test—for estimating the values of distribution parameters and for testing hypotheses about them. We found that, because of the variation from sample to sample, we run the risk of making one of two types of error when testing a hypothesis, each with quite different consequences. Normally, when testing hypotheses, we set a bound called the significance level on the probability of making a Type I error and devise our tests accordingly.

Finally, we noted the relationship between our interval estimates and our hypothesis tests.

Exercise 3.31. Make a list of all the italicized terms in this chapter. Provide a definition for each one along with an example.

Exercise 3.32. A farmer was scattering seeds in a field so they would be at least a foot apart 90% of the time. On the average, how many seeds should the farmer sow per square foot?

The answer to Exercise 3.1 is yes, of course; an observation or even a sample of observations from one population may be larger than observations from another population even if the vast majority of observations are quite the reverse. This variation from observation to observation is why before a drug is approved for marketing, its effects must be demonstrated in a large number of individuals and not just in one or two.

4

TESTING HYPOTHESES

In this chapter, we develop improved methods for testing hypotheses by means of the bootstrap, introduce parametric hypothesis testing methods, and apply these and other methods to problems involving one sample, two samples, and many samples. We then address the obvious but essential question: How do we choose the method and the statistic that is best for the problem at hand?

4.1. ONE-SAMPLE PROBLEMS

A fast-food restaurant claims that 75% of their revenue is from the "drive-thru." They've collected two weeks' worth of receipts from the restaurant and turned them over to you. Each day's receipt shows the total revenue and the drive-thru revenue for that day.

They do not claim their drive-thru produces 75% of their revenue, day in and day out, only that their overall average is 75%. In this section, we consider four methods for testing the restaurant's hypothesis.

4.1.1. Percentile Bootstrap

We've already made use of the percentile or uncorrected bootstrap on several occasions, first to estimate precision and then to obtain interval estimates for

Introduction to Statistics Through Resampling Methods and R/S-PLUS®, By Phillip I. Good
Copyright © 2005 by John Wiley & Sons, Inc.

population parameters. Readily computed, the bootstrap seems ideal for use with the drive-thru problem. Still, if something seems too good to be true, it probably is. Unless corrected, bootstrap interval estimates are *inaccurate* (i.e., they will include the true value of the unknown parameter less often than the stated confidence probability) and *imprecise* (i.e., they will include more erroneous values of the unknown parameter than is desirable). When the original samples contain less than a hundred observations, the confidence bounds based on the primitive bootstrap may vary widely from simulation to simulation.

What this means to us is that even if the hypothesized value (75%) lies outside a 95% bootstrap confidence interval, we still run the risk of making an error more than $100\% - 95\% = 5\%$ of the time in rejecting the hypothesis.

4.1.2. Parametric Bootstrap

If we know something about the population from which the sample is taken, we can improve our bootstrap confidence intervals, making them both more accurate (more likely to cover the true value of the population parameter) and more precise (narrower and thus less likely to include false values of the population parameter). For example, if we know that this population has an exponential distribution, we would use the sample mean to estimate the population mean. Then, we would draw a series of random samples of the same size as our original sample from an exponential distribution whose mathematical expectation was equal to the sample mean to obtain a confidence interval for the population parameter of interest.

This parametric approach is of particular value when we are trying to estimate one of the tail percentiles such as P_{10} or P_{90}—for the sample alone seldom has sufficient information.

```
➢ #The following R program uses an exponential
  distribution as the basis of a parametric bootstrap to
  obtain a 90% confidence interval for the IQR of the
  population from which the data set A was taken.
➢ We make use of the results of Exercise 3.10, which
  tells us that the expected value of an exponential
  distribution with parameter λ is 1/λ.
➢ #Create a vector in which to store the IQRs
➢ IQR = numeric(1000)
➢ #Set up a loop to generate the 1000 IQRs
➢ for (i in 1:1000) {
➢          bA=sample (A, n, replace=T)
➢          IQR[i]  =  IQR[i]  = qexp(.75,1/mean(bA))-
           qexp(.25,1/mean(bA))
➢          }
➢ quantile (IQR , probs = c(.05,.95))
```

There is also a parametric bootstrap function in S+Resample; for instructions on using it see `help(pbootstrap)`.

Exercise 4.1. Obtain a 90% confidence interval for the mean time to failure of a new component based on the following observations:

```
46  97  27  32  39  23  53  60  145  11  100  47  39
1  150  5  82  115  11  39  36  109  52  6  22  193
10  34  3  97  45  23  67  0  37.
```

Exercise 4.2. Would you accept or reject the hypothesis at the 10% significance level that the mean time to failure in the population from which the sample depicted in Exercise 4.1 was drawn is 97?

Exercise 4.3. Obtain an 80% confidence interval using the parametric bootstrap for the IQR of the LSAT data. *Careful*: What would be the most appropriate continuous distribution to use?

4.1.3. Student's *t*

One of the first hypothesis tests to be developed was that of Student's *t*. This test, which dates back to 1908, takes advantage of our knowledge that the distribution of the mean of a sample is usually close to that of a normal distribution. When our observations are normally distributed, then the statistic

$$t = \frac{\overline{X}. - \theta}{s/\sqrt{n}}$$

has a *t*-distribution with $n - 1$ degrees of freedom, where n is the sample size, θ is the population mean, and s is the standard deviation of the sample. Two things should be noted about this statistic:

1. Its distribution is independent of the unknown population variance.
2. If we guess wrong about the value of the unknown population mean and subtract a guesstimate of θ smaller than the correct value, then the observed values of the *t*-statistic will tend to be larger than the values predicted from a comparison with the Student's *t* distribution.

We can make use of this latter property to obtain a test of the hypothesis that the percentage of drive-in sales averages 75%, not just for our sample of sales data, but also for past and near-future sales. (Quick: Would this be a one-sided or a two-sided test?) We use R to subtract 75 from each observation in the sample, compute the t-statistic, and compare the resulting statistic with a table of the t-statistic.

Entering the R commands in the R interpreter,

```
➢ Sales = c( 80, 81, 65, 72, 73, 69, 70, 79)    #test data
➢ t.test(Sales - 75 , alternative="t")
```

yields the following output:

```
    One Sample t-test
data: Sales - 75
t = -0.67, df = 7, p-value = 0.5244
alternative hypothesis: true mean is not equal to 0
95 percent confidence interval:
 -6.227986   3.477986
sample estimates:
mean of x
  -1.375
```

We accept the claim of the restaurant's owner.

Exercise 4.4. Would you accept or reject the restaurant's hypothesis at the 5% significance level after examining the entire two weeks' worth of data: 80, 81, 65, 72, 73, 69, 70, 79, 78, 62, 65, 66, 67, 75?

Exercise 4.5. In describing the extent to which we might extrapolate from our present sample of drive-in data, we used the qualifying phrase "near-future." Is this qualification necessary or would you feel confident in extrapolating from our sample to all future sales at this particular drive-in? If not, why not?

Exercise 4.6. While some variation is expected in the width of screws coming off an assembly line, the ideal width of this particular type of screw is 10.00 and the line should be halted if it looks as if the mean width of the screws produced will exceed 10.01 or fall below 9.99. Based on the following ten observations, would you call for the line to halt so they can adjust the milling machine: 9.983, 10.020, 10.001, 9.981, 10.016, 9.992, 10.023, 9.985, 10.035, 9.960?

Exercise 4.7. In the preceding exercise, what kind of economic losses do you feel would be associated with Type I and Type II errors?

4.2. COMPARING TWO SAMPLES

In this section, we'll examine the use of the binomial, Student's t, permutation methods, and the bootstrap for comparing two samples and then address the question of which is the best test to use.

4.2.1. Comparing Two Poisson Distributions

Suppose in designing a new nuclear submarine you become concerned about the amount of radioactive exposure that will be received by the crew. You conduct a test of two possible shielding materials. During 10 minutes of exposure to a power plant using each material in turn as a shield, you record 14 counts with material A, and only four with experimental material B. Can you conclude that B is safer than A?

The answer lies not with the Poisson but the binomial. If the materials are equal in their shielding capabilities, then each of the 18 recorded counts is as likely to be obtained through the first material as through the second. In other words, under the null hypothesis you would be observing a binomial distribution with 18 trials each with probability $\frac{1}{2}$ of success or $B(18, \frac{1}{2})$.

I used just such a procedure in analyzing the results of a large-scale clinical trial involving some 100,000 service men and women who had been injected with either a new experimental vaccine or a saline control. Epidemics among service personnel can be particularly serious as they live in such close quarters. Fortunately, there were few outbreaks of the disease we were inoculating against during our testing period. Fortunate for the men and women of our armed services, that is.

When the year of our trial was completed, only 150 individuals had contracted the disease, which meant an effective sample size of 150. The differences in numbers of diseased individuals between the control and treated groups were not statistically significant.

Exercise 4.8. Can you conclude that material B is safer than A?

4.2.2. What Should We Measure?

Suppose you've got this strange notion that your college's hockey team is better than mine. We compare win/lose records for last season and see that while McGill won 11 of its 15 games, your team won only 8 of 14. But is this difference statistically significant?

With the outcome of each game being success or failure, and successive games being independent of one another, it looks at first glance as if we have two series of binomial trials (as we'll see in a moment, this is highly questionable). We could derive confidence intervals for each of the two binomial parameters. If these intervals do not overlap, then the difference in win/lose records is statistically significant. But do win/lose records really tell the story?

Let's make the comparison another way by comparing total goals. McGill scored a total of 28 goals last season and your team 32. Using the approach described in the preceding section, we could look at this set of observations as a binomial with $28 + 32 = 60$ trials, and test the hypothesis that $p \leq \frac{1}{2}$ (i.e., McGill is no more likely to have scored the goal than your team) against the alternative that $p > \frac{1}{2}$.

This latter approach has several problems. For one, your team played fewer games than McGill. But more telling, and the principal objection to all the methods we've discussed so far, the schedules of our two teams may not be comparable.

With binomial trials, the probability of success must be the same for each trial. Clearly, this is not the case here. We need to correct for the differences among opponents. After much discussion—what else is the off-season for?—you and I decide to award points for each game using the formula $S = O + GF - GA$, where GF stands for goals for, GA for goals against, and O is the value awarded for playing a specific opponent. In coming up with this formula and with the various values for O, we relied not on our knowledge of statistics but on our hockey expertise. This reliance on domain expertise is typical of most real-world applications of statistics.

The point totals we came up with read like this:

| McGill | 4, –2, 1, 3, 5, 5, 0, –1, 6, 2, 2, 3, –2, –1, 4 |
| Your school | 3, 4, 4, –3, 3, 2, 2, 2, 4, 5, 1, –2, 2, 1 |

Curiously, your school's first four point totals, all involving games against teams from other leagues, were actually losses, their high point value being the result of the high-caliber of the opponents. I'll give you guys credit for trying.

Exercise 4.9. Basing your decision on the point totals, use Student's t to test the hypothesis that McGill's hockey team is superior to your school's team. Is this a one-sided or two-sided test? If you are using R, be sure to check the options for `t.test` using the `help` command.

4.2.3. Permutation Monte Carlo

Straightforward application of the permutation methods discussed in Section 3.6.1 to the hockey data is next to impossible. Imagine how many years it would take us to look at all $\binom{14+15}{15}$ possible rearrangements! What we can do today—something not possible with the primitive calculators available in the 1930s when permutation methods were first introduced—is to look at a large random sample of rearrangements. Here's an R program that computes the original sum, then performs a *Monte Carlo*, that is, a computer simulation, to obtain the sums for 400 random rearrangements of labels:

```
➤ N = 400    #number of rearrangements to be examined
➤ sumorig = sum(McGill)
➤ n = length(McGill)
➤ cnt= 0    #zero the counter
➤ #Stick both sets of observations in a single vector
➤ A = c(McGill,Other)
➤ for (i in 1:N){
➤        D= sample (A,n)
➤        if (sum(D) <= sumorig)cnt=cnt+1
➤    }
➤ cnt/N    #one-sided p-value
```

Note that the R **sample()** instruction used here differs from the **sample()** instruction used to obtain a bootstrap in that the sampling is done *without* replacement.

Using S+Resample, we may also do:

```
➤ perm = permutationTest2(McGill, sum, data2 = Other,
➤ alternative = "less")
➤ perm # this prints the p-value, among other things
➤ plot(perm)
```

This latter instruction yields a plot that nicely illustrates the relationship between the observed value and the null distribution.

Exercise 4.10. Show that we would have got exactly the same *p*-value had we used the difference in means between the samples instead of the sum of the observations in the first sample as our permutation test statistic.

Exercise 4.11. (for mathematics and statistics majors only) Show that we would have got exactly the same *p*-value had we used the *t*-statistic as our test statistic.

Exercise 4.12. Use the Monte Carlo approach to rearrangements to test against the alternative hypothesis that McGill's hockey team is superior to your school's team.

4.2.4. One- versus Two-Sided Tests

The preceding is an example of a *one-sided test* in which we test a hypothesis "The expected value of Y is not larger than the expected value of X," against a one-sided alternative, "The expected value of Y is larger than the expected value of X." We perform the test by taking independent random samples from the X and Y populations and comparing their means. If the mean of the Y sample is less than or equal to the mean of the X sample, we accept the hypothesis. Otherwise, we examine a series of random rearrangements in order to obtain a permutation distribution of the test statistic and a p-value.

Now suppose, we state our hypothesis in the form, "The expected values of Y and X are equal," and the alternative in the form, "The expected values of Y and X are not equal." Now the alternative is two sided. Regardless of which sample mean is larger, we will need to consider a random series of rearrangements. And, as a result, we will need to double the resulting count in order to obtain a two-sided p-value.

Exercise 4.13. The following samples were taken from two different populations:

$$samp1: 1.05, 0.46, 0, -0.40, -1.09, .88$$
$$samp2: .55, .20, 1.52, 2.67$$

Were the samples drawn from two populations with the same mean?

Exercise 4.14. Modify the R code for the Monte Carlo permutation test so as to (a) increase its efficiency in the one-sided case and (b) be applicable to either one-sided or two-sided tests.

4.2.5. Bias-Corrected Nonparametric Bootstrap

The primitive bootstrap is most accurate when the observations are drawn from a normal distribution. The bias-corrected-and-accelerated bootstrap takes advantage of this as follows: Suppose θ is the parameter we are trying to estimate, $\hat{\theta}$ is the estimate, and we are able to come up with a monotone increasing transformation m such that $m(\theta)$ is normally distributed about $m(\hat{\theta})$. We could use this normal distribution and the bootstrap to obtain an unbiased confidence interval, and then apply a back-transformation to obtain an almost-unbiased confidence interval.

The good news is that we don't actually need to determine what the function m should be or to compute it to obtain the desired interval. To use R for this purpose, you'll need to download and install the "Boot" package of functions and make use of the functions **boot()** and **boot.ci()**.

Make sure you are connected to the Internet and then type

```
➤ install.packages ("boot")
```

The installation which includes downloading, unzipping, and integrating the new routines is done automatically. The installation needs to be done once and once only. But each time before you can use any of the boot library routines, you'll need to load the supporting functions into computer memory by typing.

```
➤ library (boot)
```

R imposes this requirement to keep the amount of memory its program uses to the absolute minimum.

We'll need to employ two functions from the boot library.

The first of these functions **boot (Data, Rfunction, number)** has three principal arguments. **Data** is the name of the data set you want to analyze, **number** is the number of bootstrap samples you wish to draw, and **Rfunction** is the name of an R function you must construct separately to generate and hold the values of existing R statistics functions such as **median()** or **var()** whose value you want a bootstrap interval estimate of. For example,

```
➤ f.median<-  function (y, id) {
+ median (  y [id] )
+ }
```

where R knows id will be a vector of form $1:n$. Then

```
➤ boot.ci(boot (classdata,  f.median,  400),  conf
  =  0.90)
```

will calculate a 90% confidence interval for the median of the classdata based on 400 simulations.

To obtain similar results with S-PLUS, we use the **limits.bca()** function, first creating an object containing bootstrap statistics by calling bootstrap.

```
➤ boot = bootstrap(data, median)
➤ limits.bca(boot)
```

Exercise 4.15. Compare the 90% confidence intervals for the variance of the population from which the following sample of billing data was taken for (a) the original primitive bootstrap, (b) the bias-corrected-and-accelerated nonparametric bootstrap, and (c) the parametric bootstrap assuming the billing data are normally distributed, (d) the parametric bootstrap assuming the billing data are exponentially distributed.

Hospital Billing Data
4181, 2880, 5670, 11620, 8660, 6010, 11620, 8600, 12860, 21420, 5510, 12270, 6500, 16500, 4930, 10650, 16310, 15730, 4610, 86260, 65220, 3820, 34040, 91270, 51450, 16010, 6010, 15640, 49170, 62200, 62640, 5880, 2700, 4900, 55820, 9960, 28130, 34350, 4120, 61340, 24220, 31530, 3890, 49410, 2820, 58850, 4100, 3020, 5280, 3160, 64710, 25070

Obtaining a BCa interval for a difference in means, based on samples taken from two distinct populations, is more complicated. We need make use of an R construct known as the data frame.

Suppose you've collected the following data on starting clerical salaries in Shelby County, Georgia during the periods 1975–1976 and 1982–1983.

$525 "female", $500 "female", $550 "female", $700 "male", $576 "female", $886 "male", $458 "female", $600 "female", $600 "male", $850 "male", $800 "female", $850 "female", $850 "female", $800 "female"

and would like a confidence interval for the difference in starting salaries of men and women.

First, we enter the data and store it a data frame as follows:

```
salary = c(525,500,550,700,576,886,458,600,600,850,800,
           850,850,800)
sex = c("female","female","female","male","female","male",
        "female","female","male","male","female","female",
        "female","female")
dframe = data.frame(sex,salary)
```

Then, we construct a function to calculate the difference in means between bootstrap samples of men and women:

```
fm.mndiff<- function(dframe,id){
   yvals<- dframe[[2]][id]
   mean(yvals[dframe[[1]]=="female"])-mean(yvals[dframe
   [[1]]== "male"])
   }
```

The last step, computing the BCa interval, uses the same command as in our earlier example:

```
boot.ci(boot(dframe,fm.mndiff,400),conf = 0.90)
```

4.2.6. Two-Sample *t*-test

For the same reasons that Student's *t* was an excellent choice in the one-sample case, it is recommended for comparing samples of continuous data from two populations providing that the only difference between the two is in their mean value; that is, the distribution of one is merely shifted with respect to the other so that $F_1[x] = F_2[x - \Delta]$. The test statistic is $(\overline{X}_{1.} - \overline{X}_{2.})/\hat{s}$, where $\hat{s}$ is an estimate of the *standard error* of the numerator;

$$\hat{s} = \sqrt{\frac{\sum(X_{1j} - \overline{X}_{1.})^2 / (n_1 - 1) + \sum(X_{2j} - \overline{X}_{2.})^2 / (n_2 - 1)}{n_1 + n_2 - 2}}$$

Note that the square of the *t*-statistic is the ratio of the variance *between* the samples from your school and McGill to the variance *within* these samples.

Here is an R program to make the computation:

```
➢ library(stats)
➢ McGill=c(4,-2,1,   3,5,5,0,-1,6,2,2,3,2,-1,4)
➢ YourSchool =  c(3,4,   4,-3,3,2,2,   2,   4,5,1,-2,2,   1)
➢ t.test(McGill,YourSchool, var.equal=TRUE,
   alternative="less")
```

And here are the results:

```
        Two Sample t-test
data:   McGill and YourSchool
t = -0.0726, df = 27, p-value = 0.4713
alternative hypothesis: true difference in means is less
   than 0
95 percent confidence interval:
   -Infinity 1.497741
sample estimates:
mean of x mean of y
1.933333   2.000000
```

Note that in the R instruction

```
➢ t.test(McGill,YourSchool, var.equal=TRUE,
   alternative="less")
```

we have specified that the variances of the hypothetical populations[1] from which these samples of scores are drawn are equal and that we want a one-

---

[1] The "hypothetical population" in this case consists of the scores of all the games in which these particular teams might have participated.

sided test; that is, our hypothesis is that McGill's hockey team is better than or equal to your school's team and the one-sided alternative is that it is not as good (or "less"). The resulting confidence interval is one-sided. We interpret the 95% confidence interval ($-\infty$, 1.5) as meaning that the probability that McGill is better than your school by in excess of 1.5 is less than 5%.[2]

Had our hypothesis been stated in the form McGill and Your School are equally good at hockey against the *two-sided* alternative that one school is better than the other at the sport, we'd have written the R instruction:

```
➤ t.test(McGill,YourSchool,var.equal=TRUE)
```

and obtained a two-sided confidence interval.

```
data: McGill and YourSchool
      t = -0.0726,df = 27,p-value = 0.9427
alternative hypothesis: true difference in means is not
   equal to 0
95 percent confidence interval:
-1.951197 1.817864
```

4.3. WHICH TEST SHOULD WE USE?

Four different tests were used for our two-population comparisons. Two of these were *parametric* tests that obtained their *p*-values by referring to parametric distributions such as the binomial and Student's *t*. Two were *resampling methods*—bootstrap and permutation test—that obtained their *p*-values by sampling repeatedly from the data at hand.

In some cases, the choice of test is predetermined; for example, when the observations take or can be reduced to those of a binomial distribution. In other instances, we need to look more deeply into the consequences of our choice. In particular, we need to consider the assumptions under which the test is valid, the effect of violations of these assumptions, and the Type I and Type II errors associated with each test.

4.3.1. *p*-Values and Significance Levels

In the preceding sections we have referred several times to *p*-values and significance levels. We have used both in helping us to make a decision whether to accept or reject a hypothesis and, in consequence, to take a course of action that might result in gains or losses.

---

[2] I'm not crazy about this result despite the fact it is statistically correct.

To see the distinction between the two concepts, please run the following program:

```
➢ x=rnorm(10)
➢ x
➢ t.test(x)
➢ x=rnorm(10)
➢ x
➢ t.test(x)
```

There's no misprint in the above lines of R code. What this program does is take two distinct samples, each of size 10, from a normal population with mean zero and unit variance, display the values in each instance, and provide two separate confidence intervals for the sample mean, one for each sample. The composition of the two samples varies, the value of the *t*-statistic varies, the *p-values* vary, and the boundaries of the confidence interval vary. What remains unchanged is the *significance level* of 100% − 95% = 5% that is used to make decisions.

You aren't confined to a 5% significance level. The R statement,

```
➢ t.test (x, conf.level=0.90)
```

yields a confidence level of 90% or, equivalently, a significance level of 10%.

In clinical trials of drug effectiveness, one might use a significance level of 10% in pilot studies, but would probably insist on a significance level of 1% before investing large amounts of money in further development.

In summary, *p*-values vary from sample to sample, while significance levels are fixed.

Significance levels establish limits on the overall frequency of Type I errors. The significance levels and confidence bounds of parametric and permutation tests are exact only if all the assumptions that underlie these tests are satisfied. Even when the assumptions that underlie the bootstrap are satisfied, the claimed significance levels and confidence bounds of the bootstrap are only approximations. The greater the number of observations in the original sample, the better this approximation will be.

4.3.2. Test Assumptions

Virtually all statistical procedures rely on the assumption that our observations are independent of one another. When this assumption fails, the computed *p*-values may be far from accurate, and a specific significance level cannot be guaranteed.

All statistical procedures require that at least one of the following successively stronger assumptions be satisfied under the hypothesis of no differences among the populations from which the samples are drawn:

1. The observations all come from distributions that have the same value of the parameter of interest.
2. The observations are *exchangeable*; that is, each rearrangement of labels is equally likely.
3. The observations are *identically distributed* and come from a distribution of known form.

The first assumption is the weakest. If this assumption is true, a nonparametric bootstrap test[3] will provide an exact significance level with very large samples. The observations may come from different distributions providing that they all have the same parameter of interest. In particular, the nonparametric bootstrap can be used to test whether the expected results are the same for two groups even if the observations in one of the groups are more variable than they are in the other.[4]

If the second assumption is true, the first assumption is also true. If the second assumption is true, a permutation test will provide exact significance levels even for very small samples.

The third assumption is the strongest assumption. If it is true, the first two assumptions are also true. This assumption must be true for a parametric test to provide an exact significance level.

An immediate consequence is that if observations come from a multiparameter distribution such as the normal, then all parameters, not just the one under test, must be the same for all observations under the null hypothesis. For example, a *t*-test comparing the *means* of two populations requires that the *variances* of the two populations be the same.

4.3.3. Robustness

When a test provides almost exact significance levels despite a violation of the underlying assumptions, we say that it is *robust*. Clearly, the nonparametric bootstrap is more robust than the parametric since it has fewer assumptions. Still, when the number of observations is small, the parametric bootstrap, which makes more effective use of the data, will be prefer-

---

[3] Any bootstrap but the parametric bootstrap.
[4] We need to modify our formula and our existing R program if we suspect this to be the case; see Chapter 8.

able providing enough is known about the shape of the distribution from which the observations are taken.

When the variances of the populations from which the observations are drawn are not the same, the significance level of the bootstrap is not affected. Bootstrap samples are drawn separately from each population. Small differences in the variances of two populations will leave the significance levels of permutation tests relatively unaffected but they will no longer be exact. Student's t should not be used when there are clear differences in the variances of the two groups.

On the other hand, Student's t is the exception to the rule when we say parametric tests should only be used when the distribution of the underlying observations is known. Student's t tests for differences between means, and means, as we've already noted, tend to be normally distributed even when the observations they summarize are not.

Exercise 4.16. Repeat the following 1000 times:

(a) Generate two independent samples of size 4 from a distribution that is a mixture of 0.7 $N(0,1)$ and 0.3 $N(2,1)$. Use the command:

```
Data = rnorm(8,2*rbinom(4,1,.3),1)
```

(b) Use a Monte Carlo of 400 rearrangements to derive a p-value for a test of the equality of the means of the two populations from which the samples are drawn against the one-sided alternative that the mean of the second population is larger.

(c) Use a t-test to derive a p-value for a test of the equality of the means of the two populations from which the samples are drawn against the one-sided alternative that the mean of the second population is larger.

(d) Plot the cumulative distribution functions of the resulting distributions of p-values side by side on a single chart and record the their 1st, 5th, 10th, 90th, 95th, and 99th percentiles.

4.3.4. Power of a Test Procedure

Statisticians call the probability of rejecting the null hypothesis when an alternative hypothesis is true the *power* of the test. If we were testing a food additive for possible carcinogenic (cancer-producing) effects, this would be the probability of detecting a carcinogenic effect. The power of a test equals one minus the probability of making a Type II error. The greater the power, the smaller the Type II error, the better off we are.

Power depends on all of the following:

1. The true value of the parameter being tested—the greater the gap between our primary hypothesis and the true value, the greater the power will be. In our example of a carcinogenic substance, the power of the test would depend on whether the substance was a strong or a weak carcinogen and whether its effects were readily detectable.
2. The significance level—the higher the significance level (10% rather than 5%), the larger the probability of making a Type I error we are willing to accept, and the greater the power will be. In our example, we would probably insist on a significance level of 1%.
3. The sample size—the larger the sample, the greater the power will be. In our example of a carcinogenic substance, the regulatory commission (the FDA in the United States) would probably insist on a power of 80%. We would then have to increase our sample size in order to meet their specifications.
4. The method used for testing. Obviously, we want to use the most powerful possible method.

Exercise 4.17. To test the hypothesis that consumers can't tell your cola from Coke, you administer both drinks in a blind tasting to ten people selected at random. (a) To ensure that the probability of a Type I error is just slightly more than 5%, how many people should correctly identify the glass of Coke before you reject this hypothesis? (b) What is the power of this test if the probability of an individual correctly identifying Coke is 75%?

Exercise 4.18. What is the power of the test in the preceding exercise if the probability of an individual correctly identifying Coke is 90%?

Exercise 4.19. If you test 20 people rather than 10, what will be the power of a test at the 5% significance level if the probability of correctly identifying Coke is 75%?

Exercise 4.20. Physicians evaluate diagnostic procedures on the basis of their "sensitivity" and "selectivity." Sensitivity is defined as the percentage of diseased individuals that are correctly diagnosed as such. Is sensitivity related to significance level and power? How? Selectivity, also referred to as specificity, is defined as the percentage of those diagnosed as suffering from a given disease that actually have the disease. Can selectivity be related to the concepts of significance level and power? If so, how?

Exercise 4.21. Suppose we wish to test the hypothesis that a new vaccine will be more effective than the old vaccine in preventing infectious pneumonia. We decide to inject some 1000 patients with the old vaccine and 1000 patients with the new vaccine and follow them for one year. Can we guarantee the power of the resulting hypothesis test?

Exercise 4.22. Show that the power of a test can be compared to the power of an optical lens in at least one respect.

Exercise 4.23. Repeat the following 1000 times:

(a) Generate a sample of size 4 from a distribution that is a mixture of 0.7 $N(0,1)$ and 0.3 $N(2,1)$. Generate a second independent sample of size 4 from a $N(2,1)$.

(b) Use a Monte Carlo of 400 rearrangements to derive a p-value for a test of the equality of the means of the two populations from which the samples are drawn against the one-sided alternative that the mean of the second population is larger.

(c) Use a t-test to derive a p-value for a test of the equality of the means of the two populations from which the samples are drawn against the one-sided alternative that the mean of the second population is larger.

(d) Plot the cumulative distribution functions of the resulting distributions of p-values side by side on a single chart and record the their 1st, 5th, 10th, 90th, 95th, and 99th percentiles.

4.3.5. Testing for Correlation

To see how we would go about finding the most powerful test in a specific case, consider the problem of deciding whether two variables are correlated. Let's take another look at the data from my sixth-grade classroom. The arm span and height of the five shortest students in my sixth-grade class are (139, 137), (140, 138.5), (141, 140), (142.5, 141), (143.5, 142). Both arm spans and heights are in increasing order. Is this just coincidence? Or is there a causal relationship between them or between them and a third hidden variable? What is the probability that an event like this could happen by chance alone?

The test statistic of choice is the Pitman correlation, $S = \Sigma_{i=1}^{n} a_i h_i$, where (a_k, h_k) denotes the pair of observations made on the kth individual. To prove to your own satisfaction that S will have its maximum when both arm spans and heights are in increasing order, imagine that the set of arm spans $\{a_k\}$ denotes the widths and $\{h_k\}$ the heights of a set of rectangles. The area inside the rectangles, S, will be at its maximum when the smallest width is paired with the smallest height, and so forth. If your intuition is more geometric than algebraic, prove this result by sketching the rectangles on a piece of graph paper.

We could list all possible permutations of both arm span and height along with the value of S, but this won't be necessary. We can get exactly the same result if we fix the order of one of the variables, the height, for example, and look at the 5! = 120 ways in which we could rearrange the arm span readings:

| (140, 137) | (139, 138.5) | (141, 140) | (142.5, 141) | (143.5, 142) |
| (141, 137) | (140, 138.5) | (139, 140) | (142.5, 141) | (143.5, 142) |

and so forth.

Obviously, the arrangement we started with is the most extreme, occurring exactly one time in 120 by chance alone. Applying this same test to all 22 pairs of observations, we find the odds are less than 1 in a million that what we observed occurred by chance alone and conclude that arm span and height are directly related.

```
> #Calculating a p-value for correlation via a Monte Carlo
> armspan = c(139, 140, 141, 142.5, 143.5)
> height = c (137, 138.5, 140, 141, 142)
> rho0 = cor(armspan, height)
> cnt= 0
> for (i in 1:400){
+           D = sample (armspan)
+           rho = cor(D, height)
+           # Counting correlation larger than original by
              chance
+           if (rho0 <= rho ) cnt=cnt+1
+           }
> cnt/400                        #pvalue
```

In S-PLUS we would write

```
> perm = permutationTest(armspan, cor(armspan, height),
  alternative = "greater")
> perm # This prints the p-value and other quantities
> plot(perm) # Plot the permutation distribution and
  observed correlation
```

Note that we would get exactly the same p-value if we used as our test statistic the Pearson correlation:

$$\rho = \sum_{i=1}^{n} a_i h_i / \sqrt{\mathrm{Var}[a] * \mathrm{Var}[h]}$$

This is because the variances of a and h are left unchanged by rearrangements. A rearrangement that has a large value of S will have a large value of ρ and vice versa.

Exercise 4.24. The correlation between the daily temperatures in Cairns and Brisbane is 0.29, while between Cairns and Sydney it is 0.52. Or should that be the other way around?

Exercise 4.25. Do DDT residues have a deleterious effect on the thickness of a cormorant's eggshell? (Is this a one-sided or a two-sided test?)

DDT residue in yolk (ppm)	65	98	117	122	393
Thickness of shell (mm)	0.52	0.53	0.49	0.49	0.37

Exercise 4.26. Is there a statistically significant correlation between the LSAT score and the subsequent GPA in law school?

Exercise 4.27. If we find there is a statistically significant correlation between the LSAT score and the subsequent GPA, does this mean the LSAT score of a prospective law student will be a good predictor of that student's subsequent GPA?

4.4. SUMMARY AND REVIEW

In this chapter, we derived permutation, parametric and bootstrap tests of hypothesis for a single sample, for comparing two samples, and for bivariate correlation. We showed how to improve the accuracy and precision of bootstrap confidence intervals. We explored the relationships and distinctions among p-values, significance levels, alternative hypotheses, and sample sizes. And we provided some initial guidelines to use in the selection of the appropriate test.

We showed how libraries could be used to enlarge R's capabilities: library(boot) has the function **boot.ci()** with which to perform the BCa bootstrap. We introduced an R function **cor()** for computing the correlation coefficient and showed how R's **sample()** function could be used to obtain random rearrangements of labels.

Exercise 4.28. Make a list of all the italicized terms in this chapter. Provide a definition for each one along with an example.

Exercise 4.29. Some authorities have suggested that when we estimate a *p*-value via a Monte Carlo, as in Section 4.2.3, that we should include the original observations as one of the rearrangements. Instead of reporting the *p*-value as cnt/N, we would report it as (cnt+1)/(N+1). Explain why this would give a false impression. [*Hint:* Reread Chapter 2 if necessary.]

Exercise 4.30. Efron and Tibshirani (1993) report the survival times in days for a sample of 16 mice undergoing a surgical procedure. The mice were randomly divided into two groups. The following survival times in days were recorded for a group of seven mice that received a treatment expected to prolong their survival:

➤ g.trt = c(94,197,16,38,99,141,23)

The second group of nine underwent surgery without the treatment and had these survival times in days:

➤ g.ctr = c(52,104,146,10,51,30,40,27,46)

(a) Use permutation methods to test the hypothesis that the treatment does not increase survival time.
(b) Provide a 75% confidence interval for the difference in mean survival days for the sampled population based on 1000 bootstrap samples.

Exercise 4.31. Which test would you use for a comparison of the following treated and control samples?

$$\text{control} = c(4, 6, 3, 4, 7, 6)$$

$$\text{treated} = c(14, 6, 3, 12, 7, 15).$$

5

DESIGNING AN EXPERIMENT OR SURVEY

Suppose you were a consulting statistician[1] and were given a data set to analyze. What is the first question you would ask? "What statistic should I use?" No, your first question always should be "How were these data collected?"

Experience teaches us that garbage in, garbage out or GIGO. In order to apply statistical methods, you need to be sure that samples have been drawn at random from the population(s) you want represented and are representative of those populations. You need to be sure that observations are independent of one another and that outcomes have not been influenced by the actions of the investigator or survey taker.

Many times people who consult statisticians don't know the details of the data collection process or they do know and look guilty and embarrassed when asked. All too often, you'll find yourself throwing your hands

---

[1] The idea of having a career as a consulting statistician may strike you as laughable or even distasteful. I once had a student who said he'd rather eat worms and die. Suppose then that you've eaten worms and died only to wake to discover that reincarnation is real and that to expiate your sins in the previous life you've been reborn as a consulting statistician. I'm sure that's what must have happened in my case.

Introduction to Statistics Through Resampling Methods and R/S-PLUS®, By Phillip I. Good
Copyright © 2005 by John Wiley & Sons, Inc.

in the air and saying, "if only you'd come to me to design your experiment in the first place."

The purpose of this chapter is to take you step by step through the design of an experiment and a survey. You'll learn the many ways in which an experiment can go wrong. And you'll learn the right things to do to ensure your own efforts are successful.

5.1. THE HAWTHORNE EFFECT

The original objective of the industrial engineers at the Hawthorne plant of Western Electric was to see whether a few relatively inexpensive improvements would increase workers' productivity. They painted the walls green and productivity went up. They hung posters and productivity went up. Then, just to prove how important bright paint and posters were to productivity, they removed the posters and repainted the walls a dull gray only to find that, once again, *productivity went up!*

Simply put, these industrial engineers had discovered that the mere act of paying attention to a person modifies that person's behavior. (Note, the same is true for animals.)

You've probably noticed that you respond similarly to attention from others, though not always positively. Taking a test under the watchful eye of an instructor is quite different from working out a problem set in the privacy of your room.

Physicians and witch doctors soon learn that merely giving a person a pill (any pill) or dancing a dance often results in a cure. This is called the *placebo* effect. If patients think they are going to get better, they do get better. Thus, regulatory agencies insist that before they approve a new drug, it be tested side by side with a similar looking, similar tasting *placebo*. If the new drug is to be taken twice a day in tablet form, then the placebo must also be given twice a day, also as a tablet, and not as a liquid or an injection. And, most important, the experimental subject should not be aware of which treatment she is receiving. Studies in which the treatment is concealed from the subject are known as *single-blind* studies.

The doctor's attitude is as important as the treatment. If part of the dance is omitted—a failure to shake a rattle, why bother if the patient is going to die anyway—the patient may react differently. Thus, the agencies responsible for regulating drugs and medical devices (in the United States this would be the FDA) now also insist that experiments be *double blind.* Neither the patient nor the doctor (or whoever administers the pill to the patient) should know whether the pill that is given the patient is an active drug or a placebo. If the patient searches the doctor's face for clues—will this experimental pill really help me?—she'll get the same response whether she is in the treatment group or is one of the *controls*.

Note: The double blind principle also applies to experimental animals. Dogs and primates are particularly sensitive to their handlers' attitudes.

5.1.1. Crafting an Experiment

In the very first set of clinical data that was brought to me for statistical analysis, a young surgeon described the problems he was having with his chief of surgery. "I've developed a new method for giving arteriograms which I feel can cut down on the necessity for repeated amputations. But my chief will only let me try out the technique on patients that he feels are hopeless. Will this affect my results?" It would and it did. Patients examined by the new method had a very poor recovery rate. But, of course, the only patients who'd been examined by the new method were those with a poor prognosis. The young surgeon realized that he would not be able to test his theory until he was able to assign patients to treatment at random.

Not incidentally, it took us three more tries until we got this particular experiment right. In our next attempt, the chief of surgery—Mark Craig of St. Eligius in Boston—announced that he would do the "random" assignments. He finally was persuaded to let me make the assignment using a table of random numbers. But then he announced that he, and not the younger surgeon, would perform the operations on the patients examined by the traditional method to make sure "they were done right." Of course, this turned a comparison of methods into a comparison of surgeons and intent.

In the end, we were able to create the ideal "double blind" study: The young surgeon performed all the operations, but the incision points were determined by his chief after examining one or the other of the two types of arteriogram.

Exercise 5.1. Each of the following studies is fatally flawed. Can you tell what the problem is in each instance and, as important, why it is a problem?

(a) *Class Action.* Larry the lawyer could barely pay his rent when he got the bright idea of looking through the county-by-county leukemia rates for our state. He called me a week later and asked what I thought of the leukemia rate in Kay County. I gave a low whistle. "Awfully high," I said.

   The next time I talked to Larry, he seemed happy and prosperous. He explained that he'd gone to Kay County once he'd learned that the principal employer in that area was a multinational chemical company. He'd visited all the families whose kids had come down with leukemia and signed them up for a class action suit. The company had quickly settled out of court when they looked at the figures.

"How'd you find out about Kay County?" I asked.

"Easy, I just ordered all the counties in the state by their leukemia rates and Kay came out on top."

(b) *Controls.* Danielle routinely tested new drugs for toxicity by injecting them in mice. In each case, she'd take five animals from a cage and inject them with the drug. To be on the safe side, she'd take the next five animals from the cage, inject them with a saline solution, and use them for comparison purposes.

(c) *Survey.* Reasoning, correctly, that he'd find more students home at dinnertime, Tom brought a set of survey forms back to his fraternity house and interviewed his frat brothers one by one at the dinner table.

(d) *Treatment Allocation.* Fully aware of the influence that a physician's attitude could have on a patient's recovery, Betty, a biostatistician, provided the investigators in a recent clinical trial with bottles of tablets that were labeled only A or B.

(e) *Clinical Trials.* Before a new drug can be marketed, it must go through a succession of clinical trials. The first set of trials (Phase I) is used to establish the maximum tolerated dose. They are usually limited to 25 or so test subjects who will be observed for periods of several hours to several weeks. The second set of trials (Phase II) is used to establish the minimum effective dose; they also are limited in duration and in the number of subjects involved. Only in Phase III are the trials expanded to several hundred test subjects who will be followed over a period of months or even years. Up until the 1990s, only males were used as test subjects in order to spare women the possibility of unnecessary suffering.

(f) *Comparison.* Contrary to what one would expect from the advances in medical care, there were 2.1 million deaths from all causes in the United States in 1985, compared to 1.7 million in 1960.

(g) *Survey.* The Federal Trade Commission surveyed former correspondence school students to see how they felt about the courses they had taken some two to five years earlier.[2] The survey was accompa-

---

[2] Macmillan, Inc. 96 F.T.C. 208 (1980).

nied by a form letter signed by an FTC attorney that began: "The Bureau of Consumer Protection is gathering information from those who enrolled in . . . to determine if any action is warranted." Questions were multiple choice and did not include options for "I don't know" or "I don't recall."

5.2. DESIGNING AN EXPERIMENT OR SURVEY

Before you complete a single data collection form:

1. Set forth your objectives and the use you plan to make of your research.
2. Define the population(s) to which you will apply the results of your analysis.
3. List all possible sources of variation.
4. Decide how you will cope with each source. Describe what you will measure and how you will measure it. Define the experimental unit and all end points.
5. Formulate your hypothesis and all of the associated alternatives. Define your end points. List possible experimental findings along with the conclusions you would draw and the actions you would take for each of the possible results.
6. Describe in detail how you intend to draw a representative random sample from the population.
7. Describe how you will ensure the independence of your observations.

5.2.1. Objectives

In my experience as a statistician, the people who come to consult me before they do an experiment (an all-too-small minority of my clients) aren't always clear about their objectives. I advise them to start with their reports, to write down what they would most like to see in print. For example:

"Fifteen thousand of 17,500 surveys were completed and returned. Over half of the respondents were between the ages of 47 and 56. Thirty-six percent (36%) indicated that they were currently eligible or would be eligible for retirement in the next three years. However, only 25% indicated they intended to retire in that time. Texas can anticipate some 5000 retirees in the next three years."

Or

> "Over a three-month period, 743 patients self-administered our psyllium preparation twice a day. Changes in the Klozner-Murphy self-satisfaction scale over the course of treatment were compared with those of 722 patients who self-administered an equally foul-tasting but harmless preparation over the same time period.
>
> "All patients in the study reported an increase in self-satisfaction, but the scores of those taking our preparation increased an average of 2.3 ± 0.5 points more than those in the control group.
>
> "Adverse effects included. . . .
>
> "If taken as directed by a physician, we can expect those diagnosed with . . ."

I have my clients write in exact numerical values for the anticipated outcomes—their best guesses, as these will be needed when determining sample size. My clients go over their reports several times to ensure they've included all end points and as many potential discoveries as they can—"only 25% indicated an intent to retire in that time." Once the report is fleshed out completely, they know what data needs to be collected and do not waste their time and their company's time on unnecessary or redundant effort.

Exercise 5.2. Throughout this chapter, you'll work on the design of a hypothetical experiment or survey. If you are already well along in your studies, it could be an actual one! Start now by writing the results section.

5.2.2. Sample from the Right Population

Be sure you will be sampling from the population of interest as a whole rather than from an unrepresentative subset of that population. The most famous blunder along these lines was basing the forecast of Dewey over Truman in the 1948 U.S. presidential election on a telephone survey: Those who owned a telephone and responded to the survey favored Dewey; those who voted did not.

An economic study may be flawed because we have overlooked the homeless. This was among the principal arguments the cities of New York and Los Angeles advanced against the use of the 1990 and 2000 census to determine the basis for awarding monies to cities. See *City of New York v. Dept of Commerce.*[3]

---

[3] 822 F. Supp. 906 (E.D.N.Y., 1993).

An astrophysical study was flawed because of overlooking galaxies whose central surface brightness was very low. And the FDA's former policy of permitting clinical trials to be limited to men (see Exercise 5.1e) was just plain foolish.

Plaguing many surveys are the uncooperative and the nonresponder. Invariably, follow-up surveys of these groups show substantial differences from those who responded readily the first-time around. These follow-up surveys aren't inexpensive—compare the cost of mailing out a survey to telephoning or making face-to-face contact with a nonresponder. But if one doesn't make these calls, one may get a completely unrealistic picture of how the population as a whole would respond.

Exercise 5.3. You be the judge. In each of the following cases, how would you rule?

(a) The trial of *People v. Sirhan*[4] followed the assassination of presidential candidate Robert Kennedy. The defense appealed the guilty verdict alleging the jury was a nonrepresentative sample, offering anecdotal evidence based on the population of the northern United States. The prosecution said, so what, our jury was representative of Los Angeles where the trial was held. How would you rule? Note that the Sixth Amendment to the Constitution of The United States provides that "a criminal defendant is entitled to a jury drawn from a jury panel which includes jurors residing in the geographic area where the alleged crime occurred."

(b) In *People v. Harris*,[5] a survey of trial court jury panels provided by the defense showed a significant disparity from census figures. The prosecution contended that the survey was too limited, being restricted to the Superior Courts in a single district, rather than being countywide. How would you rule?

(c) Amstar Corporation claimed that "Domino's Pizza" was too easily confused with its own use of the trademark "Domino" for sugar.[6] Amstar conducted and offered in evidence a survey of heads of households in ten cities. Domino objected to this survey pointing out that it had no stores or restaurants in eight of these cities and, in the remaining two, their outlets had been open less than three months.

---

[4] 7 Cal.3d 710, 102 Cal. Rptr. 385 (1972), *cert. denied*, 410 U.S. 947.
[5] 36 Cal.3d 36, 201 Cal. Rptr. 782 (1984), *cert. denied* 469 U.S. 965, *appeal to remand* 236 Cal. Rptr. 680, 191 Cal. App. 3d 819, *appeal after remand*, 236 Cal. Rptr. 563, 217 Cal. App. 3d 1332.
[6] *Amstar Corp. v. Domino's Pizza, Inc.*, 205 U.S.P.Q 128 (N.D. Ga. 1979), *rev'd.* 615 F. 2d 252 (5th Cir. 1980).

Domino provided a survey it had conducted in its pizza parlors and Amstar objected. How would you rule?

Exercise 5.4. Describe the population from which you plan to draw a sample in your hypothetical experiment. Is this the same population you would extend the conclusions to in your report?

THE DRUNK AND THE LAMPPOST

There's an old joke dating back to at least the turn of the previous century about the drunk whom the police officer found searching for his wallet under the lamppost. The police officer offers to help and after searching on hands and knees for fifteen minutes without success asks the inebriated gentleman just exactly where he lost his wallet. The drunk points to the opposite end of the block. "Then why were you searching over here?!"

"The light's better."

It's amazing how often measurements are made because they are convenient (inexpensive and/or quick to make) than because they are directly related to the object of the investigation. Your decisions as to what to measure and how to measure it require as much or more thought as any other aspect of your investigation.

5.2.3. Coping with Variation

As noted in the very first chapter of this text, you should begin any investigation where variation may play a role by listing all possible sources of variation—in the environment, in the observer, in the observed, and in the measuring device. Consequently, you need to have a thorough understanding of the domain—biological, psychological, or seismological—in which the inquiry is set.

Will something as simple as the time of day affect results? Body temperature and the incidence of mitosis both depend on the time of day. Retail sales and the volume of mail both depend on the day of the week. In studies of primates (including you) and hunters (tigers, mountain lions, domestic cats, dogs, wolves, etc.), the gender of the observer will make a difference.

Statisticians have found four ways for coping with individual-to-individual and observer-to-observer variation:

1. *Controlling.* Making the environment for the study—the subjects, the manner in which the treatment is administered, the manner in which the observations are obtained, the apparatus used to make the measurements, and the criteria for interpretation—as uniform and homogeneous as possible.

2. *Blocking.* A clinician might stratify the population into subgroups based on such factors as age, gender, race, and the severity of the condition and restrict subsequent comparisons to individuals who belong to the same subgroup. An agronomist would want to stratify on the basis of soil composition and environment.

3. *Measuring.* Some variables such as cholesterol level or the percentage of CO_2 in the atmosphere can take any of a broad range of values and don't lend themselves to blocking. As we show in the next chapter, statisticians have methods for correcting for the values taken by these *covariates.*

4. *Randomizing.* Randomly assign patients to treatment within each block or subgroup so that the innumerable factors that can neither be controlled nor observed directly are as likely to influence the outcome of one treatment as another.

Exercise 5.5. List all possible sources of variation for your hypothetical experiment and describe how you will cope with each one.

5.2.4. Matched Pairs

One of the best ways to eliminate a source of variation and the errors in interpretation associated with it is through the use of matched pairs. Each subject in one group is matched as closely as possible by a subject in the treatment group. If a 45-year-old black male hypertensive is given a blood-pressure lowering pill, then we give a second similarly built 45-year-old black male hypertensive a placebo.

Consider the case of a fast-food chain that is interested in assessing the effect of the introduction of a new sandwich on overall sales. To do this experiment, they designate a set of outlets in different areas—two in the inner city, two in the suburbs, two in small towns, and two located along major highways. A further matching criterion is that the overall sales for the members of each pair prior to the start of the experiment were approximately the same for the months of January through March. During the month of April, the new sandwich is put on sale at one of each pair of outlets. At the end of the month, the results are recorded for each matched pair of outlets (see Table 5.1).

Table 5.1. Sales data

Outlet	1	2	3	4	5	6	7	8
New menu	48722	28965	36581	40543	55423	38555	31778	45643
Standard	46555	28293	37453	38324	54989	35687	32000	43289

To analyze this data, we consider the 28 possible rearrangements that result from the possible exchanges of labels within each matched pair of observations.

```
➢ New = c(48722, 28965, 36581, 40543, 55423, 38555, 31778,
   45643)
➢ Standard = c(46555, 28293,  37453, 38324,  54989,
   35687,  32000, 43289)
➢ N = 400    #number of rearrangements to be examined
➢ sumorig = sum(New)
➢ n = length(New)
➢ stat = numeric (n)
➢ cnt = 0  #zero the counter
➢ for (i in 1:N){
+   which = runif(n)
+   for (j in  1:n){
+        if (which[j] < 0.5)   stat[j]=New[j]
+        else stat[j] = Standard[j]
+        }
+ if (sum(stat) <= sumorig)cnt=cnt+1
➢ }
➢ cnt/N   #one-sided p-value
[1] 0.98
```

Exercise 5.6. In the preceding example, was the correct p-value 0.98, 0.02, or 0.04?

Exercise 5.7. Did the increased sales for the new menu justify the increased cost of $1200 per location?

5.2.5. The Experimental Unit

A scientist repeatedly subjected a mouse named Harold to severe stress. She made a series of physiological measurements on Harold, recording blood pressure, cholesterol levels, and white blood cell counts both before and after stress was applied for a total of 24 observations. What was the sample size?

Another experimenter administered a known mutagen—a substance that induces mutations—into the diet of a pregnant rat. When the rat gave birth, the experimenter took a series of tissue samples from each of the seven offspring, two from each of eight body regions. What was the sample size?

In each of the preceding examples, the sample size was one. In the first example, the sole *experimental unit* was Harold. In the second example, the experimental unit was a single pregnant rat. Would stress have affected a second mouse the same way? We don't know. Would the mutagen have caused similar damage to the offspring of a different rat? We don't know. We do know there is wide variation from individual to individual in their responses to changes in the environment. With data from only a single individual in hand, I'd be reluctant to draw any conclusions about the population as a whole.

Exercise 5.8. Suppose we are testing the effect of a topical ointment on pink eye. Is each eye a separate experimental unit or each patient?

5.2.6. Formulate Your Hypotheses

In translating your study's objectives into hypotheses that are testable by statistical means, you need to satisfy all of the following:

- The hypothesis must be numeric in form and must concern the value of some population parameter. Examples: More than 50% of those registered to vote in the State of California prefer my candidate. The arithmetic average of errors in tax owed that are made by U.S. taxpayers reporting $30,000 to $50,000 in income is less than $50. The addition of vitamin E to standard cell growth medium will increase the life span of human diploid fibroblasts by no less than 30 generations. Note in these examples that we've tried to specify the population from which samples are taken as precisely as possible.
- There must be at least one meaningful numeric alternative to your hypothesis.
- It must be possible to gather data to test your hypothesis.

The statement "redheads are sexy" is not a testable hypothesis. Nor is the statement "everyone thinks redheads are sexy." Can you explain why? The statement "at least 80% of Reed College students think redheads are sexy" is a testable hypothesis.

You should also decide at the same time as you formulate your hypotheses whether the alternatives of interest are one-sided or two-sided, ordered or unordered.

Exercise 5.9. Are the following testable hypotheses? Why or why not?

(a) A large meteor hitting the Earth would dramatically increase the percentage of hydrocarbons in the atmosphere.

(b) Our candidate can be expected to receive votes in the coming election.

(c) Intelligence depends more on one's genes than on one's environment.

5.2.7. What Are You Going to Measure?

In order to formulate a hypothesis that is testable by statistical means, you need to decide on the variables you plan to measure. Perhaps your original hypothesis was that men are more intelligent than women. To put this in numerical terms requires a scale by which intelligence may be measured. Which of the many scales do you plan to use and is it really relevant to the form of intelligence you had in mind?

Be direct. To find out which drugs individuals use and in what combinations, which method would yield more accurate data: (a) a mail survey of households, (b) surveying customers as they step away from a pharmacy counter, or (c) accessing pharmacy records?

Clinical trials often make use of *surrogate response variables* that are less costly or less time-consuming to measure than the actual variable of interest. One of the earliest examples of the use of a surrogate variable was when coal miners would take a canary with them into the mine to detect a lack of oxygen well before the miners themselves fell unconscious. Today, with improved technology, they would be able to measure the concentration of oxygen directly.

The presence of HIV often serves as a surrogate for the presence of AIDS. But is HIV an appropriate surrogate? Many individuals have tested positive for HIV who do not go on to develop AIDS.[7] How shall we measure the progress of arteriosclerosis? By cholesterol levels? Angiography? Electrocardiogram? Or by cardiovascular mortality?

Exercise 5.10. Formulate your hypothesis and all of the associated alternatives for your hypothetical experiment. Decide on the variables you will measure. List possible experimental findings along with the conclusions you would draw and the actions you would take for each possible outcome. (A spreadsheet is helpful for this last.)

---

[7] A characteristic of most surrogates is that they are not one-to-one with the gold standard.

5.2.8. Random Representative Samples

Is randomization really necessary? Would it matter if you simply used the first few animals you grabbed out of the cage as controls? Or if you did all your control experiments in the morning and your innovative procedures in the afternoon? Or let one of your assistants perform the standard procedure while you performed and perfected the new technique?

A sample consisting of the first few animals to be removed from a cage will not be random because, depending on how we grab, we are more likely to select more active or more passive animals. Activity tends to be associated with higher levels of corticosteroids, and corticosteroids are associated with virtually every body function.

We've already discussed in Section 5.1.1 why a simple experiment can go astray when we *confound* competing sources of variation such as the time of day and the observer with the phenomenon that is our primary interest. As we saw in the preceding section, we can block our experiment and do the control and the innovative procedure both in the afternoon and in the morning, but we should not do one at one time and one at the other. Recommended in the present example would be to establish four different blocks (you observe in the morning, you observe in the afternoon, your assistant observes in the morning, your assistant observes in the afternoon) and to replicate the experiment separately in each block.

Samples also are taken whenever records are audited. Periodically, federal and state governments review the monetary claims made by physicians and Health Maintenance Organizations for accuracy. Examining each and every claim would be prohibitively expensive, so governments limit their audits to a sample of claims. Any systematic method of sampling, examining every tenth claim, say, would fail to achieve the desired objective. The HMO would soon learn to maintain its files in an equally systematic manner, making sure that every tenth record was error- and fraud-free. The only way to ensure honesty by all parties submitting claims is to let a sequence of random numbers determine which claims will be examined.

The same reasoning applies when we perform a survey. Let us suppose we've decided to subdivide (block) the population whose properties we are investigating into strata—males, females, city dwellers, farmers—and to draw separate samples from each stratum. Ideally, we would assign a random number to each member of the stratum and let a computer's random number generator determine which members are to be included in the sample.

By the way, if we don't *block* our population, we run the risk of obtaining a sample in which members of an important subgroup are absent or underrepresented. Recall from Section 2.2.3, that a single jury (or sample) may not be representative of the population as a whole. We can forestall this happening by deliberately drawing samples from each important subgroup.[8]

Exercise 5.11. Suppose you were to conduct a long-term health survey of our armed services personnel. What subgroups would you want to consider? Why?

Exercise 5.12. Show that the size of each subsample need not be proportional to its size in the population at large. For example, suppose your objective was to estimate the median annual household income for a specific geographic area and you were to take separate samples of households whose heads were male and female, respectively. Would it make sense to take samples of the same size from each group?

5.2.9. Treatment Allocation

If the members of a sample taken from a stratum are to be exposed to differing test conditions or treatments, then we must make sure that treatment allocation is random and that the allocation is concealed from both the investigator and the experimental subjects in so far as this is possible.

Treatment allocation cannot be left up to the investigator because of the obvious bias that would result. Having a third party label the treatments (or the treated patients) with seemingly meaningless labels such as A or B won't work either. The investigator will soon start drawing conclusions— not necessarily the correct ones—about which treatment the A's received. In clinical trials, sooner or later the code will need to be broken for a patient who is exhibiting severe symptoms that require immediate treatment. Breaking the code for one patient when the A/B method of treatment allocation is used will mean the code has been broken for all patients.

Similar objections can be made to any system of treatment allocation in which subjects are assigned on a systematic basis to one treatment regimen or the other, for example, injecting the first subject with the experimental vaccine, the next with saline, and so forth. The only safe system is one in which the assignment is made on the basis of random numbers.

---

[8] There has been much argument among legal scholars as to whether such an approach would be an appropriate or constitutional way to select juries.

5.2.10. Choosing a Random Sample

My clients often provide me with a spreadsheet containing a list of claims to be audited. Using Excel, I'll insert a new column and type =RAND() in the top cell. I'll copy this cell down the column and then SORT the entire worksheet on the basis of this column. (You'll find the SORT command in Excel's DATA menu.) The final step is to use the top 10 entries or the top 100 or whatever sample size I've specified for my audit.

Exercise 5.13. Describe the method of sampling you will use in your hypothetical experiment. If you already have the data, select the sample.

Exercise 5.14. Once again, you be the judge. The California Trial Jury Selection and Management Act[9] states that:

> It is the policy of the State of California that all persons selected for jury service shall be selected *at random* from the population of the area served by the court; that all qualified persons have an equal opportunity, in accordance with this chapter, to be considered for jury service in the state and an obligation to serve as jurors when summoned for that purpose; and that it is the responsibility of jury commissioners to manage all jury systems in an efficient, equitable, and cost-effective manner in accordance with this chapter.

In each of the following cases, decide whether this Act has been complied with:

(a) A trial judge routinely excuses physicians from jury duty because of their importance to the community.

(b) Jury panels are selected from lists of drivers compiled by the Department of Motor Vehicles.[10]

(c) A trial judge routinely excuses jurors not possessing sufficient knowledge of English.[11]

(d) A trial judge routinely excuses the "less educated" (12 or less years of formal education) or "blue collar workers."[12]

(e) A trial judge routinely excuses anyone who requests to be excused.

(f) Jury selection is usually a two- or three-stage process. At the first stage a panel is selected at random from the population. At the

---

[9] Title 3, C.C.P. Section 191.

[10] *U.S. v. Bailey*, 862 F. Supp. 277 (D. Colo. 1994) *aff'd. in part, rev'd. in part*, 76 F.3d 320, *cert. denied* 116 S. Ct. 1889.

[11] *People v. Lesara*, 206 Cal. App. 3d 1305, 254 Cal. Rptr. 417 (1988).

[12] *People v. Estrada*, 93 Cal. App. 3d 76, 155 Cal. Rptr. 731 (1979).

second stage, jurors are selected from the panel and assigned to a courtroom. In *People v. Viscotti*,[13] the issue was whether the trial court erred in taking the first 12 jurors from the panel rather than selecting 12 at random. How would you rule?

(g) A jury of 12 black males was empaneled in an area where blacks and whites were present in equal numbers.

5.2.11. Ensuring Your Observations Are Independent

Independence of the observations is essential to most statistical procedures. When observations are related—as in the analysis of multifactor designs described in the next chapter—it is essential that the residuals be independent. Any kind of dependence, even if only partial, can make the analysis suspect.

Too often, surveys take advantage of the cost savings that result from naturally occurring groups such as work sites, schools, clinics, neighborhoods, even entire towns or states. Not surprisingly, the observations within such a group are correlated. Any group or cluster of individuals who live, work, study, or pray together may fail to be representative for any or all of the following reasons:

• Shared exposure to the same physical or social environment
• Self-selection in belonging to the group
• Sharing of behaviors, ideas, or diseases among members of the group

Two events A and B are independent only if knowledge of B is *never* of value in predicting A. In statistics, two events or two observations are said to be independent if knowledge of the outcome of the one tells you nothing about the likelihood of the other. My knowledge of who won the first race will not help me predict the winner of the second. On the other hand, knowledge of the past performance of the horses in the second race would be quite helpful.

The UCLA statistics professor who serves as consultant to the California State lottery assures me that the winning numbers on successive days are completely independent of one another. Despite the obvious, the second most common pick in the California lottery are the numbers that won the previous day!

Pick two voters at random and the knowledge of one person's vote won't help me forecast the other's. But if you tell me that the second person is

---

[13] 2 Cal. 4th 1, 5 Cal. Rptr.2d 495 (1992).

the spouse of the first, then there is at least a partial dependence. (The two spouses may vote differently on specific occasions, but if one generalizes to all spouses on all occasions, the dependence is obvious.) The effect of such correlation must be accounted for by the use of the appropriate statistical procedures.[14]

The price of Coca Cola stock tomorrow does depend on the closing price today. But the change in price between today and tomorrow's closing may well be independent of the change in price between yesterday's closing and today's. When monitoring an assembly line or a measuring instrument, it is the changes from hour to hour and day to day that concern us. Change is expected and normal. It is the trends in these changes that concern us as these may indicate an underlying mutual dependence on some other hidden factors.

Exercise 5.15. Review Exercise 2.20.

5.3. HOW LARGE A SAMPLE?

Once we have mastered the technical details associated with making our observations, we are ready to launch our experiment or survey, but for one unanswered question: How large a sample should we take?

The effect of increasing sample size is best illustrated in the following series of photographs copied from `http://www.oztam.com.au/faq/#erwin`. The picture (below) is comprised of several hundred thousand tiny dots (the population).

---

[14] See, for example, Feng et al. (2001).

Now suppose we were to take successive representative samples from this population consisting of 250, 1000, and 2000 dots, respectively. They are "area probability" samples of the original picture, because the dots are distributed in proportion to their distribution in the picture. If we think of homes instead of dots, this is the sampling method used for most door-to-door surveys.

Having trouble recognizing the photo? Move back 30 inches or so from the page. When your eye stops trying to read the dots, even the smallest sample provides a recognizable picture. You would have trouble picking this woman out of a group based on the 250-dot sample. But at 1000 dots, if you squint to read the pattern of light and dark, you might recognize her. At 2000 dots, you see her more clearly; but the real improvement is between 250 and 1000—an important point. In sampling, the ability to see greater detail is a "squared function"—it takes four times as large a sample to see twice the detail. This is the strength and weakness of sample-based research. You can get the general picture cheap, but precision costs a bundle.

In our hypothetical experiment, we have a choice of using either a sample of fixed size or a sequential sampling method in which we proceed in stages, deciding at each stage whether to terminate the experiment and make a decision. In the balance of this chapter, we shall focus on methods for determining a fixed sample size, merely indicating some of the possibilities associated with sequential sampling.

5.3.1. Samples of Fixed Size

Not surprisingly, many of the factors that go into determining optimal sample size are identical with those needed to determine the power of a test (Section 4.3.3):

1. The true value of the parameter being tested. The greater the gap between our primary hypothesis and the true value, the smaller the sample needed to detect the gap.
2. The variation of the observations. The more variable the observations, the more observations we will need to detect an effect of fixed size.
3. The significance level and the power. If we fix the power against a specific alternative, then working with a higher significance level (10% rather than 5%) will require fewer observations.
4. The relative costs of the observations and of the losses associated with making Type I and Type II errors. If our measurements are expensive, then to keep the overall cost of sampling under control, we may have to accept the possibility of making Type I and Type II errors more frequently.
5. The method used for testing. Obviously, we want to use the most powerful possible method to reduce the number of observations.

The sample size that we determine by consideration of these factors is the sample size we need to end the study with. We may need to take a much larger sample to begin with in order to account for drop-outs and withdrawals, animals that escape from their cages, get mislabeled or misclassified, and so forth. Retention is a particular problem in long-term studies. In a follow-up survey conducted five years after the original, one may be able to locate as few as 20% of the original participants.

We have a choice of methods for determining the appropriate sample size. If we know how the observations are distributed, we should always take advantage of our knowledge. If we don't know the distribution exactly, but the sample is large enough that we feel confident that the statistic we are interested in has almost a normal distribution, then we should take advantage of this fact. As a method of last resort, we can run a simulation or bootstrap.

In what follows, we consider each of these methods in turn.

Known Distribution

One instance in which the distribution of the observations is always known is when there are only two possible outcomes—success or failure, a vote for our candidate or a vote against. Suppose we desire to cut down a stand of 500-year-old redwoods in order to build and sell an expensive line of patio furniture. Unfortunately, the stand is located on state property but we know a politician who we feel could be persuaded to help facilitate our purchase of the land. In fact, his agent has provided us with a rate card.

The politician is up for reelection and believes that a series of TV and radio advertisements purchased with our money could guarantee him a victory. He claims that those advertisements would guarantee him 55% of the vote. Our own advisors say he'd be lucky to get 40% of the vote without our ads and the best the TV exposure could do is give him another 5%.

We decide to take a poll. If it looks like only 40% of the voters favor the candidate, we won't give him a dime. If 46% or more of the voters already favor him, we'll pay to saturate the airwaves with his promises. We decide we can risk making a Type I error 5% of the time and a Type II error at most 10% of the time. That is, if $p = 0.40$, then the probability of rejecting the hypothesis that $p = 0.40$ should be no greater than 5%. And if $p = 0.46$, then the probability of rejecting the hypothesis that $p = 0.40$ should be at least 90%.

```
➢ #Calculate the 95% percentile of the binomial
  distribution with 10 trials and p = 0.4
➢ qbinom(.95,10,.4)
[1] 7
```

The quantile function `qbinom` always gives the next largest integer quantity. Let's see what Type I error would be associated with rejecting for values larger than 6.

```
#Calculate the probability of observing more than 6
successes for a binomial distribution with 10 trials and
p = 0.4
➢ pbinom(6,10,.4, FALSE)
[1] 0.05476188
```

Excellent, now let's see what the Type II error would be:

```
#Calculate the probability of observing 6 or less
successes for a binomial distribution with 10 trials and
p = 0.46
➢ pbinom(6,10,.46)
[1] 0.8859388
```

Much too large. Let's try a larger sample size:

```
➢ qbinom(.95,100,.4)
[1] 48
➢ pbinom(47,100,.4, FALSE)
[1] 0.06378918
```

Still a little large. The Type II error is far too large:

```
➤ pbinom(47,100,.46)
[1] 0.6191224
```

Let's try a sample size of 400:

```
➤ qbinom(.95,400,.4)
[1] 176
```

The Type I error is

```
➤ pbinom(175,400,.4, FALSE)
[1] 0.05734915
The Type II error is
➤ pbinom(175,400,.46)
[1] 0.1970185
```

We're getting close. Let's try a sample size of 800:

```
➤ qbinom(.95,800,.4)
[1] 343
➤ pbinom(342,800,.40, FALSE)
[1] 0.0525966
➤ pbinom(342,800,.46)
[1] 0.03503062 which is less than 10%
```

A poll of 800 people will give me the results I need. But why pay for that large a sample, when fewer observations will still result in the desired Type I and Type II errors?

Exercise 5.16. Find to the nearest 20 observations, the smallest sample size needed to yield a Type I error of 5% when $p = 0.40$ and a Type II error of 10% when $p = 0.46$.

Exercise 5.17. A friend of ours has a "lucky" coin that seems to come up heads every time she flips it. We examine the coin and verify that one side is marked tails. How many times should we flip the coin to test the hypothesis that it is fair so that (a) the probability of making a Type I error is no greater than 10% and (b) we have a probability of 80% of detecting a weighted coin that will come up heads 70 times out of 100 on the average?

Almost Normal Data

As noted in Chapter 3, the mean of a sample will often have an almost normal distribution similar to that depicted in Figure 1.7 even when the individual observations come from some quite different distribution. See, for example, Exercise 3.14.

In the previous section, we derived the ideal sample size more or less by trial and error. We could proceed in much the same way with normally distributed data but there is a much better way. Recall that in Exercise 3.17 we showed that the variance of the mean of n observations each with variance σ^2 was σ^2/n. If we know the cutoff value for testing the mean of a normal distribution with variance 1, we can find the cutoff value and subsequently the power of a test for the mean of a normal distribution with any variance whatever.

Let's see how. Typing

```
➢ qnorm(.95)
```

we find that the 95th percentage point of an $N(0, 1)$ distribution is

```
[1] 1.644854
```

We illustrate this result in Figure 5.1.

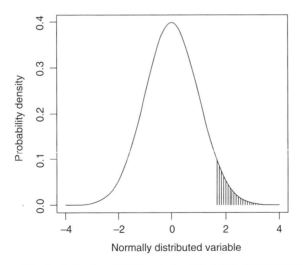

Figure 5.1. A cutoff value of 1.64 excludes 5% of $N(0, 1)$ observations.

If the true mean is actually one standard deviation larger than 0, the probability of observing a value larger than 1.644854 is given by

```
➤ pnorm(1.644854-1,0,1,FALSE)
```

or equivalently

```
1 - pnorm(1.644854-1)
[1] 0.2612401
```

We illustrate this result in Figure 5.2.

We can use Exercise 3.12 to show that if each of n independent observations is normally distributed as $N(0, 1)$ then their mean is distributed as $N(0, 1/n)$. Detecting a difference of 1 now becomes a task of detecting a difference of $\sqrt{n}$ standard deviation units. We will have power of 90% providing $1 - \text{pnorm}(1.644854 - \sqrt{n}) = 0.1$.

```
➤ qnorm(.1)
[1] -1.281552
```

Thus, we require a sample size n such that $1.644854 - \sqrt{n} = -1.281552$ or $n = 9$.

More often, we need to test against an alternative of fixed size. For example, the population mean is really equal to ten units. Thus, to determine the required sample size for testing, we would also need to know the

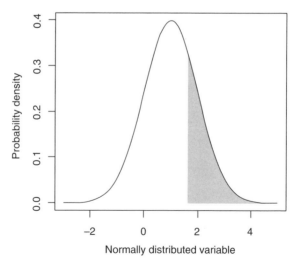

Figure 5.2. A cutoff value of 1.64 detects 26% of $N(1, 1)$ observations.

population variance or at least to have some estimate of what it is. If we have taken a preliminary sample, then the variance of this sample s^2 could serve as an estimate of the unknown population variance σ^2.

Let us suppose the sample variance is 25. The sample standard deviation is 5 and we are testing for an estimated difference of 2 standard deviations. At a significance level of 5%, we require a sample size n such that 1.644854 $- \sqrt{n} * 10/5 = -1.281552$ or no more than 3 observations.

To generalize the preceding results, suppose that C_α is the cutoff value for a test of significance at level α, and we want to have power β, to detect a difference of size δ. C_β is the value of a $N(0, 1)$ distributed variable Z for which $P\{Z > C_\beta\} = \beta$, and σ is the standard deviation of the variable we are measuring. Then $\sqrt{n} = (C_\alpha - C_\beta)\sigma/\delta$.

Exercise 5.18. How big a sample would you need to test the hypothesis that the average sixth grader is 150 cm in height at a 5% significance level so that the probability of detecting a true mean height of 160 cm is 90%? (Take advantage of the classroom data.)

Exercise 5.19. When his results were not statistically significant at the 10% level, an experimenter reported that a "new experimental treatment was ineffective." What other possibilities are there?

Bootstrap

If you have reason to believe that the distribution of the sample statistic is not normal, for example, if you are testing hypotheses regarding variances or ratios, the best approach to both power and sample size determination is to bootstrap from the empirical distribution under both the primary and the alternative hypothesis.

Recently, one of us was helping a medical device company design a comparison trial of its equipment with that of several other companies. The company had plenty of information on its own equipment but could only guess at the performance characteristics of its competitors. As they company was going to have to buy, then destroy, its competitors' equipment to perform the tests, the company wanted to keep the sample size as small as possible.

Stress test scores took values from 0 to 5 with 5 being the best. The idea was to take a sample of k units from each lot and reject if the mean score was too small. To determine the appropriate cutoff value for each prospective sample size, we took a series of simulated samples from the empirical distribution for our client's equipment. The individual frequencies for this distribution were f0, f1, f2, f3, f4, and f5. We let the computer choose a random number from 0 to 1. If this number was less than f0, we set the simulated test score to 0. If the random number was greater than f0 but less than f0 + f1, we set it to 1, and so forth. We did this k times, recorded the mean, and then repeated the entire process. For $k = 4$, 95% of the time this mean was greater than 3. So 3 was our cutoff point for $k = 4$.

Next, we guesstimated an empirical distribution for the competitor's product. We repeated the entire simulation using this guesstimated distribution. (Sometimes, you just have to reply on your best judgment.) For $k = 4$, 40% of the time the mean of our simulated samples of the competitors' products was less than 3. Not good enough. We wanted a test our product could pass and their products wouldn't.

By trial and error, we finally came up with a sample size of 6 and a test our product could pass 95% of the time and the competitors' products would fail at least 70% of the time. We were happy.

How many simulated samples n did we have to take each time? The proportion of values that fall into the rejection region is a binomial random variable with n trials and a probability β of success in each trial, where β is the desired power of the test.

We used $n = 100$ until we were ready to fine-tune the sample size, when we switched to $n = 400$.

Exercise 5.20. In preliminary trials of a new medical device, test results of 7.0 were observed in 11 out of 12 cases and 3.3 in 1 out of 12 cases. Industry guidelines specified that any population with a mean test result greater than 5 would be acceptable. A worst-case or boundary-value scenario would include one in which the test result was 7.0, three-sevenths of the time, 3.3, three-sevenths of the time, and 4.1, one-seventh of the time.

The statistical procedure with significance level 6% requires us to reject if the sample mean of subsequent test results is less than 6. What sample size is required to obtain a power of at least 80% for the worst-case scenario?

5.3.2. Sequential Sampling

The computational details of sequential sampling procedures are beyond the scope of the present text. Still, realizing that many readers will go on to design their own experiments and surveys, we devote the balance of this chapter to outlining some of the possibilities.

Stein's Two-Stage Sampling Procedure

Charles Stein's two-stage sampling procedure makes formal recognition of the need for some estimate of variation before we can decide on an optimal sample size. The procedure assumes that the test statistic will have an almost normal distribution. We begin by taking a relatively small sample and use it and the procedures of the preceding sections to estimate the optimal sample size.

If the estimated optimal sample size is less than or equal to the size of the sample we've already taken, we stop; otherwise, we take the suggested number of observations plus one.

Exercise 5.21. Apply Stein's two-stage sampling procedure to the data of Exercise 5.17. How many additional observations would we need to detect an improvement in scores of 4 units 95% of the time?

Wald Sequential Sampling

When our experiments are destructive in nature (as in testing condoms) or may have an adverse effect upon the experimental subject (as in clinical trials), we would prefer not to delay our decisions until some fixed sample size has been reached.

Figure 5.3 depicts a sequential trial of a new vaccine after eight patients who had received either the vaccine or an innocuous saline solution had

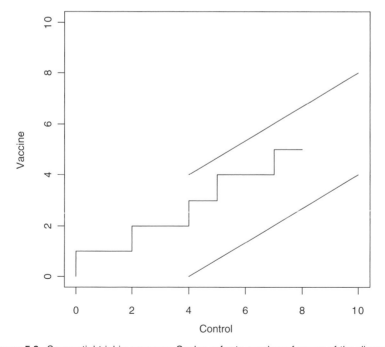

Figure 5.3. Sequential trial in progress. Scales refer to number of cases of the disease.

come down with the disease. Each time a control patient came down with the disease, the jagged line was extended to the right. Each time a patient who had received the experimental vaccine came down with the disease, the jagged line was extended upward one notch. The experiment will continue until either the jagged line crosses the lower boundary—in which case, we will stop the experiment, reject the null hypothesis, and immediately put the vaccine into production—or the jagged line crosses the upper boundary—in which case, we will stop the experiment, accept the null hypothesis, and abandon further work with this vaccine. What Abraham Wald (1945) showed in his pioneering research was that on the average the resulting sequential experiment would require many fewer observations whether or not the vaccine was effective than would a comparable experiment of fixed sample size.

Exercise 5.22. Suppose we were to take a series of observations and, after each one, reject if the test statistic were greater than the 95th percentile of its distribution under the null hypothesis. Show that the Type I error would exceed 5% even if we only took two observations.

As Exercise 5.22 illustrates, simplify performing a standard statistical test after each new observation as if the sample size were fixed will lead to inflated values of Type I error. The boundaries depicted in Figure 5.3 were obtained using formulas specific to sequential design. Not surprisingly, these formulas require us to know each and every one of the factors required to determine the number of samples when an experiment is of fixed size.

More recent developments include "group sequential designs," which involve testing not after every observation as in a fully sequential design, but rather after groups of observations (e.g., after every six months in a clinical trial). The design and analysis of such experiments is best done using specialized software such as S+SeqTrial, from http://www. insightful.com. For example, Figure 5.4 is the main menu for designing a trial to compare binomial proportions in both a treatment and a control group, with the null hypothesis being $p = 0.4$ in both groups, and the alternative hypothesis that $p = 0.45$ in the treatment group, using an "O'Brien–Fleming" design, with a total of four analyses (three "interim analyses" and a final analysis).

The resultant output (see sidebar) begins with the call to the seqDesign function that you would use if working from the command line rather than using the menu interface. The null hypothesis is that Theta [the difference in proportions (e.g., survival probability) between the two groups] is 0.0, and the alternative hypothesis is that Theta is at least 0.05. The last

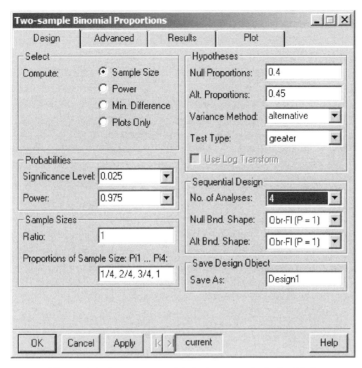

Figure 5.4. Group-sequential design menu in S+SeqTrial.

section indicates the stopping rule, which is also shown in the next plot. After 1565 observations (split roughly equally between the two groups) we should analyze the interim results. At the first analysis, if the treatment group has a survival probability that is 10% greater than the control group, we stop early and reject the null hypothesis; if the treatment group is doing 5% worse, we also stop early and accept the null hypothesis (at this point it appears that our treatment is actually killing people; there is little point in continuing the trial). Any ambiguous result, in the middle, causes us to collect more data. At the second analysis time the decision boundaries are narrower, with lower and upper boundaries 0% and 5%; stop and declare success if the treatment group is doing 5% better, stop and give up if the treatment group is doing at all worse. The decision boundaries at the third analysis time are even narrower, and at the final time (6260 total observations) they coincide; at this point we make a decision one way or the other. For comparison, the sample size and critical value for a fixed-sample trial is shown; this requires somewhat less than 6000 subjects.

```
*** Two-Sample Binomial Proportions Trial ***
Call:
seqDesign(prob.model = "proportions", arms = 2,
null.hypothesis
    = 0.4, alt.hypothesis = 0.45, ratio = c(1., 1.),
    nbr.analyses
    = 4, test.type = "greater", power = 0.975, alpha =
    0.025, beta
    = 0.975, epsilon = c(0., 1.), display.scale =
    seqScale(
  scaleType = "X"))
```

PROBABILITY MODEL and HYPOTHESES:
  Two arm study of binary response variable
  Theta is difference in probabilities (Treatment -
Comparison)
  One-sided hypothesis test of a greater alternative:
        Null hypothesis : Theta < = 0 (size = 0.025)
    Alternative hypothesis : Theta > = 0.05 (power = 0.975)
    (Emerson & Fleming (1989) symmetric test)

STOPPING BOUNDARIES: Sample Mean scale
                                    a   d
    Time 1 (N = 1565.05) -0.0500 0.1000
    Time 2 (N = 3130.09)   0.0000 0.0500
    Time 3 (N = 4695.14)   0.0167 0.0333
    Time 4 (N = 6260.18)   0.0250 0.0250

Figure 5.5 depicts the boundaries of a group sequential trial: at each of four analysis times; at each time a difference in proportions below the lower boundary or above the upper boundary causes the trial to stop; anything in the middle causes it to continue. For comparison, a fixed trial (in which one only analyzes the data at the completion of the study) is shown; this would require just under 6000 subjects for the same Type I error and power.

The major benefit of sequential designs is that we may stop early if results clearly favor one or the other hypothesis. For example, if the treatment really is worse than the control, we are likely to hit one of the lower boundaries early. If the treatment is much better than the control, we are likely to hit an upper boundary early. Even if the true difference is right in the middle between our two hypotheses, say, that the treatment is 2.5% better (when the alternative hypothesis is that it is 5% better), we may stop early on occasion. Figure 5.6 shows the average sample size as a function of Theta, the true difference in means. When Theta is less than 0% or greater

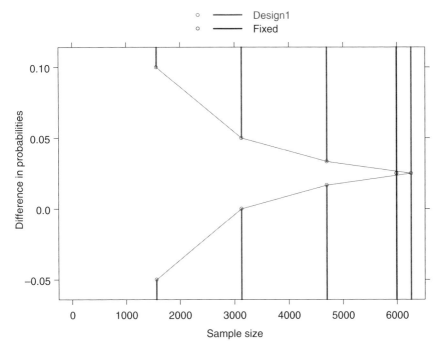

Figure 5.5. Group-sequential decision boundaries.

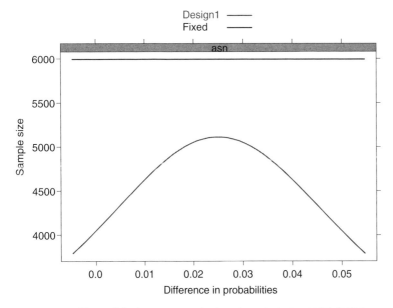

Figure 5.6. Average sample sizes for group-sequential design.

than 5%, we need about 4000 observations on average before stopping. Even when the true difference is right in the middle, we stop after about 5000 observations, on average. In contrast, the fixed-sample design requires nearly 6000 observations for the same Type I error and power.

Adaptive Sampling

The adaptive method of sequential sampling is used primarily in clinical trials where the treatment or the condition being treated presents substantial risks to the experimental subjects. Suppose, for example, 100 patients have been treated, 50 with the old drug and 50 with the new. If on reviewing the results, it appears that the new experimental treatment offers substantial benefits over the old, we might change the proportions given each treatment, so that in the next group of 100 patients, just 25 randomly chosen patients receive the old drug and 75 receive the new.

5.4. META-ANALYSIS

Such is the uncertain nature of funding for scientific investigation, that experimenters often lack the means necessary to pursue a promising line of research. A review of the literature in your chosen field is certain to turn up several studies in which the results are inconclusive. An experiment or survey has ended with results that are "almost" significant, say, with $p = 0.075$ but not $p = 0.049$. The question arises whether one could combine the results of several such studies, thereby obtaining, in effect, a larger sample size and a greater likelihood of reaching a definitive conclusion. The answer is yes, through a technique called *meta-analysis*.

Unfortunately, a complete description of this method is beyond the scope of this text. There are some restrictions on meta-analysis, for example, that the experiments whose p-values are to be combined should be comparable in nature.

Exercise 5.23. List all the respects in which you feel experiments ought be comparable in order that their p-values should be combined in a meta-analysis.

5.5. SUMMARY AND REVIEW

In this chapter, you learned the principles underlying the design and conduct of experiments and surveys. You learned how to cope with variation through controlling, blocking, measuring, or randomizing with respect to all contributing factors. You learned the importance of giving a precise explicit formulation to your objectives and hypotheses. You learned a variety of techniques to ensure that your samples would be both random and representative of the population of interest. And you learned a variety of methods for determining the appropriate sample size.

You also learned that there is much more to statistics than can be presented within the confines of a single introductory text.

Exercise 5.24. A highly virulent disease is known to affect one in 5000 people. A new vaccine promises to cut this rate in half. Suppose we were to do an experiment in which we vaccinated a large number of people, half with an ineffective saline solution and half with the new vaccine. How many people would we need to vaccinate to ensure that the probability was 80% of detecting a vaccine as effective as this one purported to be while the risk of making a Type I error was no more than 5%? [*Hint*: See Section 4.2.1.]

There was good news and bad news in just such a series of clinical trials recently. The good news was that almost none of the subjects—control or vaccine treated—came down with the disease. The bad news was that with so few diseased individuals the trials were inconclusive.

Exercise 5.25. To compare teaching methods, 20 school children were randomly assigned to one of two groups. The following are the test results:

| Conventional | 85 79 80 70 61 85 98 80 86 75 |
| New | 90 98 73 74 84 81 98 90 82 88 |

Are the two teaching methods equivalent in result? What sample size would be required to detect an improvement in scores of 5 units 90% of the time where our test is carried out at the 5% significance level?

Exercise 5.26. To compare teaching methods, 10 school children were divided into two groups. In one, they were first taught by conventional methods, tested, and then taught related but different material by an entirely new approach. The second group were taught by the new approach first. The following are the test results:

| Conventional | 85 79 80 70 61 85 98 80 86 75 |
| New | 90 98 73 74 84 81 98 90 82 88 |

Are the two teaching methods equivalent in result? What sample size would be required to detect an improvement in scores of 5 units 90% of the time? Again, the significance level for the hypothesis test is 5%.

Exercise 5.27. Make a list of all the italicized terms in this chapter. Provide a definition for each one along with an example.

6

ANALYZING COMPLEX EXPERIMENTS

In this chapter, you'll learn how to analyze a variety of different types of experimental data including changes measured in percentages, samples drawn from more than two populations, categorical data presented in the form of contingency tables, samples with unequal variances, and multiple end points.

6.1. CHANGES MEASURED IN PERCENTAGES

In the previous chapter, we learned how we could eliminate one component of variation by using each subject as its own control. But what if we are measuring weight gain or weight loss where the changes, typically, are best expressed as percentages rather than absolute values? A 250-pound person might shed 20 pounds without anyone noticing; not so with a 125-pound person.

The obvious solution is to work not with the before–after differences but with the before/after ratios.

But what if the original observations are on growth processes—the size of a tumor or the size of a bacterial colony—and vary by several orders of

Introduction to Statistics Through Resampling Methods and R/S-PLUS®, By Phillip I. Good
Copyright © 2005 by John Wiley & Sons, Inc.

magnitude? H. E. Renis of the Upjohn Company observed the following vaginal virus titers in mice 144 hours after inoculation with herpes virus type II:

Saline controls	10,000	3,000	2,600	2,400	1,500
Treated with antibiotic	9,000	1,700	1,100	360	1

In this experiment the observed values vary from 1, which may be written as 10^0, to 10,000, which may be written as 10^4 or 10 times itself 4 times. With such wide variation, how can we possibly detect a treatment effect?

The trick employed by statisticians is to use the *logarithms* of the observations in the calculations rather than their original values. The logarithm or log of 10 is 1, the log of 10,000 written log(10000) is 4. Log (0.1) is −1. (Yes, the trick is simply to count the number of decimal places that follow the leading digit.)

Using logarithms with growth and percentage-change data has a second advantage. In some instances it equalizes the variances of the observations or their ratios so that they all have the identical distribution up to a shift. Recall that equal variances are necessary if we are to apply any of the methods we learned for detecting differences in the means of populations.

Exercise 6.1. Was the antibiotic used by H. E. Renis effective in reducing viral growth? [*Hint*: First convert all the observations to their logarithms using the R function `log()`.]

Exercise 6.2. While crop yield improved considerably this year on many of the plots treated with the new fertilizer, there were some notable exceptions. The recorded after/before ratios of yields on the various plots were as follows: 2, 4, 0.5, 1, 5.7, 7, 1.5, 2.2. Is there a statistically significant improvement?

6.2. COMPARING MORE THAN TWO SAMPLES

The comparison of more than two samples is an easy generalization of the method we used for comparing two samples. As in Chapter 4, we want a test statistic that takes more or less random values when there are no differences among the populations from which the samples are taken but tends to be large when there are differences. Suppose we have taken samples of sizes $n_1, n_2, \ldots, n_I$ from I populations. Consider either of the statistics

$$F_2 = \sum_{i=1}^{I} n_i (\overline{X}_{i\cdot} - \overline{X}_{\cdot\cdot})^2$$

or

$$F_1 = \sum_{i=1}^{I} n_i |\overline{X}_{i\cdot} - \overline{X}_{\cdot\cdot}|$$

where $\overline{X}_{i\cdot}$ is the mean of the ith sample and $\overline{X}_{\cdot\cdot}$ is the grand mean of all the observations.

Recall from Chapter 1 that the symbol Σ stands for sum of, so that

$$\sum_{i=1}^{I} n_i (\overline{X}_{i\cdot} - \overline{X}_{\cdot\cdot})^2 = n_1 (\overline{X}_{1\cdot} - \overline{X}_{\cdot\cdot})^2 + n_2 (\overline{X}_{2\cdot} - \overline{X}_{\cdot\cdot})^2 + \cdots + n_I (\overline{X}_{I\cdot} - \overline{X}_{\cdot\cdot})^2$$

If the means of the I populations are approximately the same, then changing the labels on the various observations will not make any difference as to the expected value of F_2 or F_1, as all the sample means will still have more or less the same magnitude. On the other hand, if the values in the first population are much larger than the values in the other populations, then our test statistic can only get smaller if we start rearranging the observations among the samples. We can show this by drawing a series of figures as we did in Section 4.3.4 when we developed a test for correlation.

Our permutation test consists of rejecting the hypothesis of no difference among the populations when the original value of F_2 (or of F_1 should we decide to use it as our test statistic) is larger than all but a small fraction, say, 5%, of the possible values obtained by rearranging labels.

6.2.1. Programming the Multisample Comparison in R

To program the analysis of more than two populations in R, we need to program an R function. We also make use of some new R syntax to sum only those data values that satisfy a certain condition, sum(data[condition]), as in the expression term=mean(data[start:end])

```
➤ F1=function(size,data){
➤ #Size is a vector containing the sample sizes
➤ #Data is a vector containing all the data in the same
    order as the sample sizes
+     stat=0
```

```
+    start=1
+ grandmean = mean(data)
+ for (i in 1:length(size)){
+    groupMean = mean(data[seq(from = start, length =
     size[i])])
+    stat = stat + abs(groupMean - grandMean)
+    start = start + size[i]
+    }
+ return(stat)
+}
```

We use this function repeatedly in the following R program:

```
➤ # One-way analysis of unordered data via a Monte Carlo
➤ size = c(4,6,5,4)
➤ data = c(28,23,14.27, 33, 36,34, 29, 31, 34,
  18,21,20,22,24,11,14,11,16)
➤
➤ # Insert the definition of function F1 here
➤ f1=F1(size,data)
➤ #Number N of simulations determines precision of
  p-value
➤ N = 1600
➤ cnt = 0
➤ for (i in 1:N){
+ pdata = sample (data)
+ f1p=F1(size,pdata)
+ # Counting number of rearrangements for which F1 greater
    than or equal to original
+ if (f1 <= f1p) cnt=cnt+1
+ }
➤ cnt/N
```

To test the program just created, generate 16 random values with the command

```
➤ data = rnorm(16, 2*rbinom(16, 1,  .4))
```

The values generated by this command come from a mixture of normal distributions $N(0, 1)$ and $N(2, 1)$. Set `size=c(4,4,3,5)`. Now do the following exercises.

Exercise 6.3. Run your program to see if it will detect differences in these four samples despite them all being drawn from the same population.

Exercise 6.4. Modify your program by adding the following command after you've generated the data:

```
➤ data = data + c(2,2,2,2,  0,0,0,0,  0,0,0,  0,0,0,0,0)
```

Now test for differences among the populations.

Exercise 6.5. We saw in the preceding exercise that if the expected value of the first population was much larger than the expected values of the other populations that we would have a high probability of detecting the difference. Would the same be true if the mean of the second population was much higher than that of the first? Why?

Exercise 6.6. Modify your program by adding the following command after you've generated the data:

```
➤ data  =  data  +  c(1,1,1,1,     0,0,0,0,   -1.2,-1.2,-1.2,
  0 0,0,0,0)
```

Now test for differences among the populations.

Exercise 6.7. To test your command of R programming, write a function to compute F_2 and repeat the previous three exercises.

6.2.2. Reusing Your R Functions*

After you've created the function **F1** you can save it to your hard disk with the R commands

```
➤ dir.create("/Rfuncs")   # if the directory does not yet
  exist
➤ save(F1, file="/Rfuncs/pak1.Rfuncs")
```

When you need to recall this function for use at a later time, just enter the command

```
➤ load("/Rfuncs")
```

6.2.3. What Is the Alternative?

We saw in the preceding exercises that we can detect differences among several populations if the expected value of one population is much larger than the others or if the mean of one of the populations is a little higher and the mean of a second population is a little lower than the grand mean.

Suppose we represent the expectations of the various populations as follows: $EX_i = \mu + \delta_i$, where μ (pronounced mu) is the grand mean of all the populations and δ_i represents the deviation of the expected value of the ith population from this grand mean. The sum of these deviations $\Sigma\delta_i = \delta_1 + \delta_2 + \ldots + \delta_I = 0$. We will sometimes represent the individual observations in the form $X_{ij} = \mu + \delta_i + z_{ij}$, where z_{ij} is a random deviation with expected value 0 at each level of i. The permutation tests we describe in this section are applicable only if all the z_{ij} have the same distribution at each level of i.

One can show, though the mathematics is tedious, that the power of a test using the statistic F_2 is an increasing function of $\Sigma\delta^2_i$. The power of a test using the statistic F_1 is an increasing function of $\Sigma|\delta_i|$. The problem with these omnibus tests is that while they allow us to detect any of a large number of alternatives, they are not especially powerful for detecting any specific alternative. As we shall see in the next section, if we have some advance information that the alternative is, for example, an ordered dose response, then we can develop a much more powerful statistical test specific to that alternative.

Exercise 6.8. Suppose a car manufacturer receives four sets of screws, each from a different supplier. Each set is a population. The mean of the first set is 4 mm, the second set 3.8 mm, the third set 4.1 mm, and the fourth set 4.1 mm, also. What would the values of μ, δ_1, δ_2, δ_3, and δ_4 be? What would be the value of $\Sigma|\delta_i|$?

6.2.4. Testing for a Dose Response or Other Ordered Alternative

Frank, Trzos, and Good (1977) studied the increase in chromosome abnormalities and micronuclei as the dose of various compounds known to cause mutations was increased. Their object was to develop an inexpensive but sensitive biochemical test for mutagenicity that would be able to detect even marginal effects. The results of their experiment are reproduced in Table 6.1.

To analyze such data, Pitman (1937) proposes a test for linear correlation with three or more *ordered samples* using as test statistic $S = \Sigma g[i]s_i$ where s_i is the sum of the observations in the ith dose group, and $g[i]$ is any monotone increasing function of i The simplest example of such a function is $g[i] = i$, with test statistic $S = \Sigma g[i]s_i$. In this instance, based on the recommendation of experts in toxicology, we take $g[dose] = log[dose + 1]$, as the anticipated effect is proportional to the logarithm of the dose. Our test statistic is $S = \Sigma log[dose_i + 1]s_i$.

Table 6.1. Micronuclei in polychromatophilic erythrocytes and chromosome alterations in the bone marrow of mice treated with CY

Dose (mg/kg)	Number of Animals	Micronuclei per 200 cells	Breaks per 25 cells
0	4	0 0 0 0	0 1 1 2
5	5	1 1 1 4 5	0 1 2 3 5
20	4	0 0 0 4	3 5 7 7
80	5	2 3 5 11 20	6 7 8 9 9

The original data for breaks may be written in the form

$$0\ 1\ 1\ 2 \quad 0\ 1\ 2\ 3\ 5 \quad 3\ 5\ 7\ 7 \quad 6\ 7\ 8\ 9\ 9$$

As $\log(0 + 1) = 0$, the value of the Pitman statistic for the original data is $0 + 11\log(6) + 22\log(21) + 39\log(81) = 112.1$. The only larger values are associated with the small handful of rearrangements of the form

$$
\begin{array}{llll}
0\ 0\ 1\ 2 & 1\ 1\ 2\ 3\ 5 & 3\ 5\ 7\ 7 & 6\ 7\ 8\ 9\ 9 \\
0\ 0\ 1\ 1 & 1\ 2\ 2\ 3\ 5 & 3\ 5\ 7\ 7 & 6\ 7\ 8\ 9\ 9 \\
0\ 0\ 1\ 1 & 1\ 2\ 2\ 3\ 3 & 5\ 5\ 7\ 7 & 6\ 7\ 8\ 9\ 9 \\
0\ 0\ 1\ 2 & 1\ 1\ 2\ 3\ 3 & 5\ 5\ 7\ 7 & 6\ 7\ 8\ 9\ 9 \\
0\ 1\ 1\ 2 & 0\ 1\ 2\ 3\ 3 & 5\ 5\ 7\ 7 & 6\ 7\ 8\ 9\ 9 \\
0\ 1\ 1\ 2 & 0\ 1\ 2\ 3\ 5 & 3\ 5\ 6\ 7 & 7\ 7\ 8\ 9\ 9 \\
0\ 0\ 1\ 2 & 1\ 1\ 2\ 3\ 5 & 3\ 5\ 6\ 7 & 7\ 7\ 8\ 9\ 9 \\
0\ 0\ 1\ 1 & 1\ 2\ 2\ 3\ 5 & 3\ 5\ 6\ 7 & 7\ 7\ 8\ 9\ 9 \\
0\ 0\ 1\ 1 & 1\ 2\ 2\ 3\ 3 & 5\ 5\ 6\ 7 & 7\ 7\ 8\ 9\ 9 \\
0\ 0\ 1\ 2 & 1\ 1\ 2\ 3\ 3 & 5\ 5\ 6\ 7 & 7\ 7\ 8\ 9\ 9 \\
0\ 1\ 1\ 2 & 0\ 1\ 2\ 3\ 3 & 5\ 5\ 6\ 7 & 7\ 7\ 8\ 9\ 9 \\
\end{array}
$$

As there are

$$\binom{18}{4\ 5\ 4} = 771{,}891{,}120$$

rearrangements in all,[1] a statistically significant ordered dose response of $p < 0.001$ has been detected. The micronuclei also exhibit a statistically significant dose response when we calculate the permutation distribution of $S = \Sigma\log(\text{dose}_i + 1)n_i$. To make the calculations for this second test, we took advantage of the following R program:

[1] See Section 2.2.1.

```
#Defaults to N = 400 unless N is given another value
➢ pitcor.test= function(data, weights, N=400){
+  s0= sum(data*weights)
+  ct= 0
+  for(ii in 1:N){
+      dx= sample(data)
+      sx= sum(dx*weights)
+      if(sx >= s0) ct= ct+1
+      }
+  ct/N
+  }
➢ mnuc= c(0,0,0,0,1,1,1,4,5,0,0,0,4,2,3,5,11,20)
➢ dose= c(rep(0,4),rep(5,5),rep(20,4),rep(80,5))
➢ p.mnuc= pitcor.test(data=mnuc, weights=log(dose+1),
   N=1000)
➢ p.mnuc
[1] 0.005
```

A word of caution: If we use as the weights some function of the dose other than $g(\text{dose}) = \log(\text{dose} + 1)$, we might observe a different result. Our choice of a test statistic must always make practical as well as statistical sense.

Exercise 6.9. Using the data for micronuclei; see if you can detect a significant dose effect. [*Hint*: I usually use $N = 400$ repetitions to begin with. Try with both $N = 400$ and $N = 1600$.]

---

k-SAMPLE TEST FOR ORDERED SAMPLES

Hypothesis H: All distributions and all population means are the same.
Alternative K: The population means are ordered.

Assumptions under the null hypothesis:

1. Labels on the observations can be exchanged if the hypothesis is true.
2. All the observations in the *i*th sample come from the same distribution G_i, where

$$G[x] = \Pr\{X \leq x\} = F[x - \delta]$$

Test statistic

$$S = \Sigma g[i]x_{i\cdot}.$$

where $x_{i\cdot}$ is the sum of the observations in the *i*th sample.

Exercise 6.10. Aflatoxin is a common and undesirable contaminant of peanut butter. Are there significant differences in aflatoxin levels among the following brands?

Snoopy	0.5	7.3	1.1	2.7	5.5	4.3
Quick	2.5	1.8	3.6	5.2	1.2	0.7
Mrs. Good's	3.3	1.5	0.4	4.8	2.2	1.0

[*Hint*: What is the null hypothesis? What alternative or alternatives are of interest?]

Exercise 6.11. Does the amount of potash in the soil affect the strength of fibers made of cotton grown in that soil? Consider the data in the following table:

	Potash Level (lb/acre)				
	144	108	72	54	36
Breaking strength	7.46	7.17	7.76	8.14	7.63
	7.68	7.57	7.73	8.15	8.00
	7.21	7.80	7.74	7.87	7.93

6.3. EQUALIZING VARIANCES

Suppose that to cut costs on our weight loss experiment, we have each participant weigh him or herself. Some individuals will make very *precise* measurements, perhaps repeating the procedure three or four times to make sure they've performed the measurement correctly. Others, will say "close enough" and get the task done as quickly as possible. The problem with our present statistical methods is they treat each observation as if it were equally important. Ideally, we should give the least consideration to the most variable measurements and the greatest consideration to those that are least variable. The problem is that we seldom have any precise notion of what these variances are.

One possible solution is to put all results on a pass–fail or success–failure basis. This way, if the goal is to lose at least 7% of body weight, losses of 5% and 20% would offset each other, rather than a single 20% loss being treated as if it were equivalent to four losses of 5%. These pass–fail results would follow a binomial distribution and the appropriate method for testing binomial hypotheses could be applied.

The down side is the loss of information, but if our measurements are not equally precise, perhaps it is noise rather than useful information that we are discarding.

Here is a second example: An experimenter administered three drugs in random order to each of five recipients. He recorded their responses and now wishes to decide if there are significant differences among treatments. The problem is that the five subjects have quite different baselines. A partial solution would be to subtract an individual's baseline value from all subsequent observations made on that individual. But who is to say that an individual with a high baseline value will respond in the same way as an individual with a low baseline reading?

An alternate solution would be to treat each individual's readings as a *block* (see Section 5.2.3) and then combine the results. But then we run the risk that the results from an individual with unusually large responses might mask the responses of the others. Or, suppose the measurements actually had been made by five different experimenters using five different measuring devices in five different laboratories. Would it really be appropriate to combine them?

No, for sets of observations measured on different scales are not *exchangeable*. By converting the data to *ranks*, separately for each case, we are able to put all the observations on a common scale, and then combine the results.

When we replace observations by their ranks, we generally give the smallest value rank 1, the next smallest rank 2, and so forth. If there are N observations, the largest will have rank N. In Tables 6.2a and 6.2b, we've made just such a transformation.

When three readings are made on each of five subjects, there are a total of $(3!)^5 = 7776$ possible rearrangements of labels within blocks (subjects). For a test of the null hypothesis against any and all alternatives using F_2 as our test statistic, as can be seen in Table 6.2b, only $2 \times 5 = 10$ of them, a handful of the total number of rearrangements, are as or more extreme than our original set of ranks.

Table 6.2a. Original observations

	A	B	C	D	E
Control	89.7	75	105	94	80
Treatment 1	86.2	74	95	98	79
Treatment 2	76.5	68	94	93	84

Table 6.2b. Ranks

	A	B	C	D	E
Control	1	1	1	2	2
Treatment 1	2	2	2	1	3
Treatment 2	3	3	3	3	1

Exercise 6.12. Suppose we discover that the measurement for subject D for treatment 1 was recorded incorrectly and should actually be 90. How would this affect the significance of the results depicted in Table 6.2b?

Exercise 6.13. Does fertilizer help improve crop yield when there is minimal sunlight? Here are the results of a recent experiment exactly as they were recorded (numbers in bushels). No fertilizer: 5, 10, 8, 6, 9, 122; with fertilizer: 11, 18, 15, 14, 21, 14.

WHEN TO USE RANKS

1. When one or more extreme-valued observations are suspect.
2. When the methods used to make the measurements were not the same for each observation.

Many older textbooks advocate the use of tests based on ranks for a broad variety of applications. But *rank tests* are simply permutation tests applied to the ranks of observations rather than to their original values. Their value has diminished as a result of improvements in computer technology and they should not be employed except in the two instances outlined above.

Ranks are readily obtained in R by use of the function **rank**().

6.4. CATEGORICAL DATA

We have shown in two examples (Sections 4.3.4 and 6.2.3) how one may test for the independence of metric variables using a correlation coefficient. But what if our observations are categorical involving race or gender or some other categorical attribute?

Suppose on examining the cancer registry in a hospital, we uncover the following data (Table 6.3) that we put in the form of a 2×2 contingency table.

Table 6.3. Cancer survival

	Survived	Died	Total
Men	9	1	10
Women	4	10	14
Total	13	11	24

Table 6.4a. Cancer survival

	Survived	Died	Total
Men	10	0	10
Women	3	11	14
Total	13	11	24

The 9 denotes the number of males who survived, the 1 denotes the number of males who died, and so forth. The four marginal totals or *marginals* are 10, 14, 13, and 11. The total number of men in the study is 10, while 14 denotes the total number of women, and so forth.

The marginals in Table 6.3 are fixed because, indisputably, there are 11 dead bodies among the 24 persons in the study and 14 women. Suppose that before completing the table, we only have a list of the patients who survived and those who did not, but can not tell without further investigation which patients are men and which are women. Now suppose I were to hand you 10 labels with the word "man" and 14 labels with the word "woman." Under the null hypothesis, you are allowed to distribute these labels to the patients independently of their survival status.

There are a total of $\binom{24}{10}$ ways you could hand out the gender labels "man" and "woman." Of these assignments, $\binom{14}{10}\binom{10}{1}$ result in tables that are as extreme as our original table (i.e., in which 90% of the men survive) and $\binom{14}{11}\binom{10}{0}$ result in tables that are more extreme (100% of the men survive). See Tables 6.4a and 6.4b.[2]

---

[2] Note that in terms of the relative survival rates of the two sexes, Table 6.4a is more extreme than our original Table 6.2a; Table 6.4b is less extreme.

Table 6.4b. Cancer survival

	Survived	Died	Total
Men	8	2	10
Women	5	9	14
Total	13	11	24

This is a very small fraction of the total, less than 1%. To show this explicitly, we will need the aid of two R functions that we program ourselves: one that will compute factorials and one that will compute combinations. The first of these takes advantage of our knowledge that $n! = n(n-1)!$ and uses a programming technique called *recursion*:

```
➢   fact=function(n){
➢   # This assumes, without checking, that n is an integer
+        if (n==1|n==0)1
+        else n*fact(n-1)
+        }
```

We then write $\binom{n}{t}$ in the form

```
➢ comb = function (n,t) fact(n)/(fact(t)*fact(n-t))
```

We want to compute

```
➢ (10*comb(14,10) + comb(14,11))/comb(24,10)
[1] 0.0056142594
```

In S-PLUS, we can use the built-in **choose()** function:

```
➢ (10*choose(14,10) + choose(14,11))/choose(24,10)
```

From the result, 0.006, we conclude that a difference in survival rates of the two sexes at least as extreme as the difference we observed in Tables 6.2a and 6.2b is very unlikely to have occurred by chance alone. We reject the hypothesis that the survival rates for the two sexes are the same and accept the alternative hypothesis that, in this instance at least, males are more likely to profit from treatment.

6.4.1. One-Sided Fisher's Exact Test

The preceding test is known as Fisher's Exact Test as it was first described by R. A. Fisher in 1935. Before we can write a program to perform this test we need to consider the general case depicted in Table 6.5.

If the two attributes represented by the four categories are independent of one another, then each of the tables with the marginals n, m, and t is equally likely. If t is the smallest marginal, there are a total of $\binom{m+n}{t}$ possible tables. If $t - x$ is the value in the cell with the fewest observations, then

$$\sum_{k=0}^{t-x} \binom{m}{t-k}\binom{n}{k}$$

tables are as or more extreme than the one we observed.

To simplify the programming, we will assume that the smallest marginal is in the first row and the smallest cell frequency is located in the first column so that we load the data as shown:

```
➢ data = c(t-x, x, n - (t-x), m-x)
```

or, since we are more likely to have the actual cell frequencies,

```
➢ data =c(f11, f12, f21, f22)
➢ m = data[2] + data[4]
➢ n = data [1] + data [3]
➢ t = data[1] + data[2]
➢ ntab=0
➢ for (k in 0:data[1]) ntab = ntab + comb(m,t-k)*comb(n,k)
➢ ntab/comb(m+n,t)          #prints the p-value for Fisher's
  Exact    Test
```

Exercise 6.14. What is the probability of observing Table 6.3 or one more extreme by chance alone?

Table 6.5.

	Category 1	Category 2	Total
Category A	$t - x$	x	t
Category B	$n - (t - x)$	$M - x$	$M + n - t$
Total	n	m	$m + n$

Exercise 6.15. A physician has noticed that half her patients who suffer from sore throats get better within a week if they get plenty of bed rest. (At least they don't call her back to complain that they aren't better.) She decides to do a more formal study and contacts each of 20 such patients during the first week after they've come to see her. What she learns surprises her. Twelve of her patients didn't get much of any bed rest or if they did go to bed on a Monday, they were back at work on a Tuesday. Of these noncompliant patients, six had no symptoms by the end of the week. The remaining eight patients all managed to get at least three days bed rest (some by only going to work half-days) and of these, six also had no symptoms by the end of the week. Does bed rest really make a difference?

6.4.2. The Two-Sided Test

In the example of the cancer registry, we tested the hypothesis that survival rates do not depend on gender against the alternative that men diagnosed with cancer are likely to live longer than women similarly diagnosed. We rejected the null hypothesis because only a small fraction of the possible tables were as extreme as the one we observed initially. This is an example of a one-tailed test. But is it the correct test? Is this really the alternative hypothesis we would have proposed if we had not already seen the data? Wouldn't we have been just as likely to reject the null hypothesis that men and women profit the same from treatment if we had observed Table 6.6?

Of course, we would! In determining the significance level in this example, we must perform a two-sided test and add together the total number of tables that lie in either of the two extremes or tails of the permutation distribution.

Unfortunately, it is not as obvious which tables should be included in the second tail. Is Table 6.6 as extreme as Table 6.3 in the sense that it favors an alternative more than the null hypothesis? One solution is simply to double the p-value we obtained for a one-tailed test. Alternately, we can

Table 6.6.

	Survived	Died	Total
Men	0	10	10
Women	13	1	14
Total	13	11	24

Table 6.7a.

	Survived	Died	Total
Men	1	9	10
Women	12	2	14
Total	13	11	24

Table 6.7b.

	Survived	Died	Total
Men	2	8	10
Women	11	3	14
Total	13	11	24

define and use a test statistic as a basis of comparison. One commonly used measure is the Pearson χ^2 (chi-square) statistic defined for the two-by-two (2×2) contingency table after eliminating terms that are invariant under permutations as $[x - tm/(m + n)]2$. For Table 6.3, this statistic is 12.84; for Table 6.6, it is 29.34.

Exercise 6.16. Show that Table 6.7a is more extreme (in the sense of having a larger value of the chi-square statistic) than Table 6.3, but Table 6.7b is not.

6.4.3. Multinomial Tables

It is possible to extend the approach described in the previous sections to tables with multiple categories such as Tables 6.8 and 6.9.

Table 6.8.

	Full Recovery	Partial Recovery	No Improvement
Untreated			
Low dose			
High dose			

Table 6.9.

	Urban	Suburban	Rural
Republican			
Democrat			
Independent			

As in the preceding sections, we need only to determine the total number of tables with the same marginals as well as the number of tables that are as or more extreme than the table at hand. Two problems arise. First, what do we mean by more extreme? In Table 6.8, would a row that held one more case of "Full Recovery" be more extreme than a table that held two more cases of "Partial Recovery"? At least a half dozen different statistics including the Pearson χ^2 statistic have been suggested for use with tables like Table 6.9 in which neither category is ordered.

The second problem that arises lies in the computations, which are not a simple generalization of the program for the 2×2 case. Fortunately, software is commercially available.

Exercise 6.17. In a two-by-two contingency table, once we fix the marginals, we are only free to modify a single entry. In a three-by-three table, how many different entries are we free to change without changing the marginals? Suppose the table has R rows and C columns, how many different entries are we free to change without changing the marginals?

6.5. MULTIVARIATE ANALYSIS

No competent physician would make a diagnosis on the basis of a single symptom. "Got a sore throat have you? Must be strep. Here take a pill." To the contrary, she is likely to ask you to open your mouth and say "ah," take your temperature, listen to your chest and heartbeat, and perhaps take a throat culture. Similarly, a multivariate approach is far more likely to reveal differences brought about by treatment or environment, than focusing on changes in a single end point.

Will a new drug help mitigate the effects of senility? To answer this question, we might compare treated and control patients on the basis of their scores on three quizzes: current events, arithmetic, and picture completion.

Will pretreatment of seeds increase the percentage of germination, shorten the time to harvest, and increase the yield?

Will a new fuel additive improve gas mileage and ride quality in both stop-and-go and highway situations?

In trying to predict whether a given insurance firm can avoid bankruptcy, should we measure the values and ratios of agents' balances, total assets, stocks less cost, stocks less market value, expenses paid, and net premiums written?

In Section 6.5.2, we will learn how to combine the p-values obtained from separate but dependent single-variable (univariate) tests into an overall multivariate p-value.

6.5.1. Manipulating Multivariate Data in R*

To take advantage of multivariate statistical procedures, we need to maintain all the observations on a single experimental unit as a single indivisible vector. If we are to do a permutation test, labels must be applied to and exchanged among these vectors and not among the individual observations. For example, if we are studying the effects of pretreatment on seed, the data for each plot (the number of seeds that germinated, the time to the initial harvest, and the ultimate yield) must be treated as a single indivisible unit.

Data for the different variables may have been entered in separate vectors:

```
➢ germ = c(11,9,8,4)
➢ time = c(5, 4.5, 6, 8)
```

Or, the observations on each experimental unit may have been entered separately:

```
➢ one =c(11,5)
➢   two =c (9,4.5)
➢   thr = c(8,6)
➢   four = c(4,8)
```

In the first case, we put the data in matrix form as follows:

```
➢ temp  = c(germ,time)
➢ data  = matrix(temp,2,4, byrow=T)
➢ data
     [,1] [,2] [,3] [,4]
[1,] 11   9.0   8    4
[2,]  5   4.5   6    8
```

If the observations on each experimental unit have been entered separately, we would create a matrix in the following way:

```
➢ temp = c(one,two,thr, four)
➢ data = matrix (temp, 2,4)
➢ data
   [,1]  [,2]  [,3]  [,4]
[1,] 11   9.0   8    4
[2,]  5   4.5   6    8
```

We can rank all the observations on a variable-by-variable basis with the commands

```
➢ R1 = rank(data[1,])
➢ R2 = rank(data[2,])
➢ rdata = matrix(c(R1,R2),2,4,byrow=T)
```

Here is a useful R function for generating random rearrangements of multivariate data:

```
➢ rearrange =function (data){
    n=length(data[1, ])
    m= length (data[ ,1])
    vect = c(1:n)
    rvect = sample (vect)
    new = data
    for (j in 1:n){
        ind = rvect [j]
        for (k in 1:m)  new[k,j]= data [k,ind]
        }
  return(new)
  }
```

6.5.2. Pesarin–Fisher Omnibus Statistic

Let **Data** denote the original $N \times M$ matrix of multivariate observations. N, the number of rows in the matrix, corresponds to the number of experimental units. M, the number of columns in this matrix, corresponds to the number of different variables.

Our first task is to replace the matrix by a row vector $\mathbf{T_0}$ of M univariate statistics. That is, calculating a column at a time, we replace that column with a single value—a t-statistic, a mean or some other value.

To fix ideas, suppose we wish to compare two lots of fireworks and have taken a sample from each. For each firework, we have recorded whether or not it exploded, how long it burned, and some measure of its brightness. The first column of the **Data** matrix consists of a series of ones corresponding to whether or not the firework for that row exploded. Such observations are best evaluated by the methods of Section 6.4, so we replace the column by a single number—that of the value of the Pearson chi-square statistic.

The times that are recorded in the second column have approximately an exponential distribution (see Section 4.4). So unless the observation is missing (as it would be if the firework did not explode), we take its logarithm and replace the column by the mean of the observations in the first sample.

The brightness values in the third column have approximately a normal distribution, so we replace this column by the corresponding value of the t-statistic.

Our second task is to rearrange the observations in **Data** between the two samples, using the methods of the previous section so as to be sure to keep all the observations on a given experimental unit together. Again, we compute a row vector of univariate statistics **T**. We repeat this resampling process K times. If there really isn't any difference between the two lots of fireworks, then each of the vectors $\mathbf{T}_k$ obtained after rearranging the data as well as the orginal vector $\mathbf{T}_0$ are equally likely.

We would like to combine these values but have the same problem we encountered and resolved in Section 6.3: The p-values have all been measured on different scales. The solution as it was in Section 6.3 is to replace the $K + 1$ different values of each univariate test statistic by their ranks. We rank the $K + 1$ values of the chi-square statistic separately from the $K + 1$ values of the t-statistic, and so forth, so that we have m sets of ranks ranging from 1 to $K + 1$.

The ranks should be related to the extent to which the statistic favors the alternative. For example, if large values of the t-statistic are to be expected when the null hypothesis is false, then the largest value of this statistic should receive rank $K + 1$.

Our third task is to combine the ranks of the M individual univariate tests using Fisher's omnibus statistic

$$U_i = \sum_{m=1}^{M} \log\left(\frac{K+0.5-R_{im}}{K+1}\right), \quad i = 1, \ldots, K+1$$

Our fourth and final task is to determine the p-value associated with this statistic. We require three sets of computations. To simplify writing them down, we will make repeated use of an indicator function I. $I[y] = 1$ if its argument y is true; $I[y] = 0$ otherwise. For example, $I[T > 3]$ is 1 if $T > 3$ and is 0 otherwise.

1. Determine from the individual distributions (parametric or permutation) the marginal significance level of each of the single-variable statistics for the original nonrearranged observations:

$$p_m = \frac{0.5 + \sum_{k=1}^{K} I[T_{km} \geq T_{0m}]}{K+1}, \quad m = 1, \ldots, M$$

2. Combine these values into a single statistic,

$$U_0 = \sum_{m=1}^{M} \log(p_m) = -\sum_{m=1}^{M} \log\left(\frac{K + 0.5 - R_{0m}}{K+1}\right)$$

Note that R_{0k} can take any value in the range 1 to $K + 1$.

3. Determine the significance level of the combined test:

$$p = \frac{0.5 + \sum_{k=1}^{K} I[U_k \geq U_0]}{K+1}$$

6.5.3. Programming Guidelines

In order to complete the exercises in the balance of this chapter, you will need to write a program in R to perform the Pesarin–Fisher omnibus multivariate test. The program is a long one and it is essential that you proceed in a systematic fashion. Here are some guidelines.

- Envision the program in sections. For example:
    1. Enter data and store it in matrix form.
    2. Compute univariate statistics.
    3. Repeatedly resample from data and compute univariate statistics.
    4. Convert univariate statistics to ranks.
    5. Compute the Pesarin–Fisher statistic and determine its p-value.
- Write general, but test simple. Use global constants for sample sizes and number of resamples so you can make changes at a single location.
- Program and test your program a section at a time.
- Insert a comment before each line of your program describing what that line does and why. This last suggestion may seem silly and unnecessary, but suppose you need to use and modify your program a

month or a year from today. Will you still know what each line of your program is supposed to do?

- Define and use functions. For example, by defining a function to compute all your univariate statistics (each of which requires its own function), you can use that function first to compute the univariate statistics for the data with its original labeling and then for each rearrangement.
- Think first, program second. Organize your program first on paper and then transfer it section by section to the computer, filling in the line-by-line details and testing as you go. Save your work.
- Anticipate and program for exceptions. For example:

```
if (n<0 ||" k<0 || (n-trunc(n)) !=0 || (k-trunc(k))
!=0) stop ("n and k must be positive integers")
else fact(n) / (fact (k) *fact (n-k))
```

Testing and debugging a program also requires that you proceed in a systematic fashion.

- Use artificial data sets to begin with. Use the smallest possible number of samples, variables, and experimental units; for example, two samples each with two experimental units, and two observations on each experimental unit. Use distinct samples such as (1, 2) and (0, 0).
- Test a section at a time. When testing the section in which you resample, test with $N = 1$ and then $N = 2$ before increasing N to a more practical value. If the results don't make sense, print out intermediate values. For example, after you enter the data

```
data = matrix (temp, 2,4)
```

display it

```
data
```

Once you are sure your program works correctly, you can remove these redundant commands.

- Vary your test data a step at a time. For example, use three observations per sample instead of two. Next, use unequal sample sizes. Then, use three observations (variables) per experimental unit instead of two.
- Complete your tests using real data.

Exercise 6.18.   You are studying a new tranquilizer that you hope will minimize the effects of stress. The peak effects of stress manifest themselves between five and ten minutes after the stressful incident, depending on the individual. To be on the safe side, you've made observations at both the five- and ten-minute marks.

Subject	Prestress	5-minute	10-minute	Treatment
A	9.3	11.7	10.5	Brand A
B	8.4	10.0	10.5	Brand A
C	7.8	10.4	9.0	Brand A
D	7.5	9.2	9.0	New drug
E	8.9	9.5	10.2	New drug
F	8.3	9.5	9.5	New drug

How would you correct for the prestress readings? Is this a univariate or a multivariate problem? List possible univariate and multivariate test statistics. Perform the permutation tests and compare the results.

Exercise 6.19.   To help understand the sources of insolvency, Trieschman and Pinches (1973) compared the financial ratios of solvent and financially distressed insurance firms. A partial listing of their findings is included in the following table. Are the differences statistically significant? Be sure to state the specific hypotheses and alternative hypotheses you are testing. Will you use Student's t as your univariate test statistic each time? Or the difference in mean values? Or simply the sum of the observations in the first sample?

	Solvent Companies					Insolvent Companies			
	V1	V2	V3	V4		V1	V2	V3	V4
1	0.056	0.398	1.138	0.109	9	0.059	1.168	1.145	0.732
2	0.064	0.757	1.005	0.085	10	0.054	0.699	1.052	0.052
3	0.033	0.851	1.002	0.118	11	0.168	0.845	0.997	0.093
4	0.025	0.895	0.999	0.057	12	0.057	0.592	0	0.057
5	0.050	0.928	1.206	0.191	13	0.337	0.898	1.033	0.088
6	0.060	1.581	1.008	0.146	14	0.230	1	1.157	0.088
7	0.015	0.382	1.002	0.141	15	0.107	0.925	0.984	0.247
8	0.079	0.979	0.996	0.192	16	0.193	1.120	1.058	0.502

V1 = (agents balances)/(total assets)
V2 = (stocks − cost)/(stocks − market value)
V3 = (bonds − cost)/(bonds − market value)
V4 = (expenses paid)/(net premiums written)

Exercise 6.20. You wish to test whether a new fuel additive improves gas mileage and ride quality in both stop-and-go and highway situations. Taking 12 vehicles, you run them first on a highway-style track and record the gas mileage and driver's comments. You then repeat on a stop-and-go track. You empty the fuel tanks and refill, this time including the additive, and again run the vehicles on the two tracks.

The following data was supplied in part by the Stata Corporation. Use Pesarin's combination method to test whether the additive affects either gas mileage or ride quality.

Id	bmpg1	ampg	rqi1	bmpg2	ampg2	rqi2
1	20	24	0	19	23.5	1
2	23	25	0	22	24.5	1
3	21	21	1	20	20.5	0
4	25	22	0	24	20.5	−1
5	18	23	1	17	22.5	1
6	17	18	−1	16	16.5	−1
7	18	17	0	17	16.5	0
8	24	28	1	23	27.5	0
9	20	24	0	19	23.5	1
10	24	27	0	22	25.5	0
11	23	21	0	22	20.5	0
12	19	23	1	18	22.5	1

bmpg1	track 1 before additive
ampg1	track 1 after additive
rqi1	ride quality improvement track 1
bmpg2	track 2 before additive
ampg2	track 2 after additive
rqi2	ride quality improvement track 2

Exercise 6.21. The following data is based on samples taken from swimming areas off the coast of Milazzo (Italy) from April through September 1998. Included in this data set are levels of total coliform, fecal coliform, streptococci, oxygen, and temperature.

(a) Are there significant differences in each of these variables from month to month?
(b) Are there significant multivariate differences from month to month in terms of the Pesarin–Fisher omnibus statistic?
(c) Was the use of a multivariate approach of value in this example?

Month = c(4, 5, 6, 7, 8, 9, 4, 5, 6, 7, 8, 9, 4, 5, 6, 7, 8, 9, 4, 5, 6, 7, 8, 9, 4, 5, 6, 7,
8, 9, 4, 5, 6, 7, 8, 9, 4, 5, 6, 7, 8, 9, 4, 5, 6, 7, 8, 9, 4, 5, 6, 7, 8, 9, 4, 5, 6,
7, 8, 9, 4, 5, 6, 7, 8, 9, 4, 5, 6, 7, 8, 9, 4, 5, 6, 7, 8, 9)

Temp = c(14, 17, 24, 21, 22, 20, 14, 17, 24, 21, 23, 22, 14, 17, 25, 21, 21, 22, 14,
17, 25, 21, 25, 20, 14, 17, 25, 21, 21, 19, 14, 17, 25, 21, 25, 19, 14, 16,
25, 21, 25, 19, 15, 19, 18, 21, 25, 19, 15, 19, 18, 20, 22, 17, 15, 19, 18,
20, 23, 18, 15, 17, 18, 20, 25, 17, 15, 17, 18, 20, 25, 19, 15, 17, 19, 20,
25, 18, 15, 18, 19, 21, 24, 19)

TotColi = c(30, 22, 16, 18, 32, 40, 50, 34, 32, 32, 34, 18, 16, 19, 65, 54, 32, 59,
45, 27, 88, 32, 78, 45, 68, 14, 54, 22, 25, 32, 22, 17, 87, 17, 46, 23, 10,
19, 38, 22, 12, 26, 8, 8, 11, 19, 45, 78, 6, 9, 87, 6, 23, 28, 0, 0, 43, 8, 23,
19, 0, 5, 28, 19, 14, 32, 12, 17, 33, 21, 18, 5, 22, 13, 19, 27, 30, 28, 16,
6, 21, 27, 58, 45)

FecColi = c(16, 8, 8, 11, 11, 21, 34, 11, 11, 7, 11, 6, 8, 6, 35, 18, 18, 21, 13, 9, 32,
11, 29, 11, 28, 7, 12, 7, 12, 9, 10, 3, 43, 5, 12, 14, 4, 9, 8, 10, 4, 12, 0,
4, 7, 5, 12, 26, 0, 3, 32, 0, 8, 12, 0, 0, 21, 0, 7, 8, 0, 0, 17, 4, 0, 14, 0, 0,
11, 7, 6, 0, 8, 0, 0, 6, 4, 5, 10, 14, 3, 8, 12, 11, 27)

Strep = c(8, 4, 3, 5, 6, 9, 18, 7, 6, 3, 5, 2, 6, 0, 11, 2, 12, 7, 9, 3, 8, 6, 9, 4, 8, 2, 7,
3, 5, 3, 4, 0, 16, 0, 6, 5, 0, 0, 2, 4, 0, 4, 0, 0, 3, 0, 5, 14, 0, 0, 11, 0, 2, 4, 0,
0, 3, 0, 2, 2, 0, 0, 7, 2, 0, 5, 0, 0, 7, 4, 3, 0, 2, 0, 2, 2, 3, 5, 8, 1, 2, 4, 5, 7)

O$_2$ = c(95.64, 102.09, 104.76, 106.98, 102.6, 109.15, 96.12, 111.98, 100.67,
103.87, 107.57, 106.55, 89.21, 100.65, 100.54, 102.98, 98, 106.86, 98.17,
100.98, 99.78, 100.87, 97.25, 97.78, 99.24, 104.32, 101.21, 102.73, 99.17,
104.88, 97.13, 102.43, 99.87, 100.89, 99.43, 99.5, 99.07, 105.32, 102.89,
102.67, 106.04, 106.67, 98.14, 100.65, 103.98, 100.34, 98.27, 105.69, 96.22,
102.87, 103.98, 102.76, 107.54, 104.13, 98.74, 101.12, 104.98, 101.43,
106.42, 107.99, 95.89, 104.87, 104.98, 100.89, 109.39, 98.17, 99.14, 103.87,
103.87, 102.89, 108.78, 107.73, 97.34, 105.32, 101.87, 100.78, 98.21, 97.66,
96.22, 22, 99.78, 101.54, 100.53, 109.86)

6.6. SUMMARY AND REVIEW

In this chapter, you learned to analyze a variety of different types of experimental data. You learned to convert your data to logarithms when changes would be measured in percentages or when analyzing data from dividing populations. You learned to convert your data to ranks when observations were measured on different scales or when you wanted to minimize the importance of extreme observations.

You learned to specify in advance of examining your data whether your alternative hypotheses of interest were one-sided or two-sided, ordered or unordered.

You learned how to compare binomial populations and to analyze 2×2 contingency tables. You learned how to combine the results of multiple end points and to use the combination to compare samples from two or more populations.

You were introduced to built-in R functions `abs()`, `cov()`, and `ranks()` and created several new R functions of your own. You learned how to store data in a matrix and techniques for the systematic development and testing of lengthy programs.

Exercise 6.22. Write definitions for all italicized words in this chapter.

Exercise 6.23. Many surveys employ a nine-point Likert scale to measure respondent's attitudes where a value of "1" means the respondent definitely disagrees with a statement and a value of "9" means they definitely agree. Suppose you have collected the views of Republicans, Democrats, and Independents in just such a form. How would you analyze the results?

7

DEVELOPING MODELS

In this chapter you will learn valuable techniques with which to develop forecasts and classification schemes. These techniques have been used to forecast parts sales by the Honda Motors Company and epidemics at naval training centers, to develop criteria for retention of marine recruits, optimal tariffs for Federal Express, and multitiered pricing plans for Delta Airlines. And these are just examples in which I've been personally involved!

7.1. MODELS

A model in statistics is simply a way of expressing a quantitative relationship between one variable, usually referred to as the *dependent variable*, and one or more other variables, often referred to as the *predictors*. We began our text with a reference to Boyle's Law for the behavior of perfect gases, $V = KT/P$. In this version of Boyle's Law, V (the volume of the gas) is the dependent variable; T (the temperature of the gas) and P (the pressure exerted on and by the gas) are the predictors; and K (known as Boyle's constant) is the *coefficient* of the ratio T/P.

An even more familiar relationship is that between the distance S traveled in t hours and the velocity V of the vehicle in which we are traveling:

Introduction to Statistics Through Resampling Methods and R/S-PLUS®, By Phillip I. Good
Copyright © 2005 by John Wiley & Sons, Inc.

$S = Vt$. Here S is the dependent variable and V and t are predictors. If we travel at a velocity of 60 mph for three hours, we can plot the distance we travel over time, using the following R code:

```
➢  Time = c(0.5,1,1.5,2, 2.5, 3)
➢  S = 60*Time
➢  plot(S, xlab="Time")
```

I attempted to drive at 60 mph on a nearby highway recently past where a truck had recently overturned with the following results:

```
➢  SReal = c(32,66,75,90,115,150)
➢  plot(Time,S, ylab="Distance")
➢  points(Time,SReal,pch=19)
```

As you can see, the reality was quite different from the projected result. My average velocity over the three-hour period was equal to distance traveled/time = 150/3 = 50 miles per hour. Or $S_i = 50 * \text{Time}_i + z_i$, where the z_i are random deviations from the expected distance.

Let's compare the new model with the reality:

```
➢  S = 50*Time
➢  SReal = c(32,66,75,90,115,150)
➢  plot(Time,S, ylab="Distance")
➢  points(Time,SReal,pch=19)
```

Our new model, depicted in Figure 7.1, is a much better fit, agreed?

7.1.1. Why Build Models?

We develop models for at least three different purposes. First, as the term "predictors" suggests, models can be used for *prediction*. A manufacturer of automobile parts will want to predict parts sales several months in advance to ensure its dealers have the necessary parts on hand. Too few parts in stock will reduce profits, too many may necessitate interim borrowing. So entire departments are hard at work trying to come up with the needed formula.

(At one time, I was part of just such a study team. We soon realized that the primary predictor of parts sales was the weather. Snow, sleet, and freezing rain sent sales skyrocketing. Unfortunately, predicting the weather is as or more difficult than predicting parts sales.)

Models can be used to develop additional insight into cause-and-effect relationships. At one time, it was assumed that the growth of the welfare

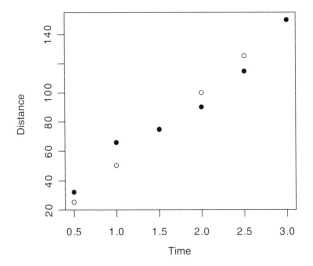

Figure 7.1. Distance expected at 50 mph versus distance observed.

case load L was a simple function of time t, so that $L = ct$, where the growth rate c was a function of population size. Throughout the 1960s, in state after state, the constant c constantly had to be adjusted upward if this model were to fit the data. An alternative and better fitting model proved to be $L = ct + dt^2$, an equation often used in modeling the growth of an epidemic. As it proved, the basis for the new second-order model was the same as it was for an epidemic: Welfare recipients were spreading the news of welfare availability to others who had not yet taken advantage of the program much as diseased individuals might spread an infection.

Boyle's Law seems to fit the data in the sense that if we measure both the pressure and volume of gases at various temperatures, we find that a plot of pressure times volume versus temperature yields a straight line. Or if we fix the volume, say, by confining all the gas in a chamber of fixed size with a piston on top to keep the gas from escaping, a plot of the pressure exerted on the piston against the temperature of the gas yields a straight line.

Observations such as these both suggested and confirmed what is known today as kinetic molecular theory.

A third use for models is in *classification*. At first glance, the problem of classification might seem quite similar to that of prediction. Instead of predicting that Y would be 5 or 6 or even 6.5, we need only predict that Y will be greater or less than 6, for example. But the loss functions for the two problems are quite different. The loss connected with predicting y_p when the observed value is y_o is usually a monotone increasing function of the difference between the two. By contrast, the loss function connected with a classification problem has jumps, being zero if the classification is correct, and taking one of several possible values otherwise, depending on the nature of the misclassification.

Not surprisingly, different modeling methods have developed to meet the different purposes. In the balance of the chapter, we shall consider three primary modeling methods: linear regression whose objective is to predict the expected value of a given dependent variable, quantile regression whose objective is to predict one or more quantiles of the dependent variable's distribution, and CART, which may be used both for classification and as an alternative to linear regression.

7.1.2. Caveats

The modeling techniques that you learn in this chapter may seem impressive—they require extensive calculations that only a computer can do—so that I feel it necessary to issue three warnings.

- You cannot use the same data both to formulate a model and to test it. It must be independently validated.
- A cause-and-effect basis is required for every model, just as molecular theory serves as the casual basis for Boyle's Law.
- Don't let your software do your thinking for you. Just because a model fits the data does not mean that it is appropriate or correct. Again, it must be independently validated and have a cause-and-effect basis.

You may have heard that having a black cat cross your path will bring bad luck. Don't step in front of a moving vehicle to avoid that black cat unless you have some causal basis for believing that black cats can affect your luck. (And why not white cats or tortoise shell?) I avoid cats myself because cats lick themselves and shed their fur; when I breathe cat hairs, the traces of saliva on the cat fur trigger an allergic reaction that results in the blood vessels in my nose dilating. Now, that is a causal connection.

7.2. REGRESSION

Regression combines two ideas with which we gained familiarity in previous chapters:

1. Correlation or dependence among variables.
2. Additive model.

Here is an example. Anyone familiar with the restaurant business (or indeed, with any number of businesses that provide direct service to the public, including the Post Office) knows that the volume of business is a function of the day of the week. Using an *additive model*, we can represent business volume via the formula

$$V_{ij} = \mu + \delta_i + z_{ij}$$

where V_{ij} is the volume of business on the ith day of the jth week, μ is the average volume, δ_i is the deviation from the average volume observed on the ith day of the week, $i = 1, \ldots, 7$, and the z_{ij} are independent, identically distributed random fluctuations.

Many physiological processes such as body temperature have a circadian rhythm, rising and falling each 24 hours. We could represent temperature by the formula

$$T_{ij} = \mu + \delta_i + z_{ij}$$

where i (in minutes) takes values from 1 to $24*60$, but this would force us to keep track of 1441 different parameters. Besides, we can get almost as good a fit to the data using the formula

$$E(T_{tj}) = \mu + \beta\cos(2\pi(t + 300)/1440) \tag{7.1}$$

If you are not familiar with the cos( ) function, you can use R to gain familiarity as follows:

```
➤  hour=c(0:24)
➤  P=cos(2*pi*(hour+6)/24)
➤  plot(P)
```

Note how the cos( ) function first falls then rises, undergoing a complete cycle in a 24-hour period.

Why use a formula as complicated as Equation 7.1? Because now we have only two parameters we need to estimate, μ and β. For predicting body temperature $\mu = 98.6$ and $\beta = 0.4$ might be reasonable choices, though, of course, the values of these parameters will vary from individual to individual. For me, $\mu = 97.6$.

Exercise 7.1. If $E(Y) = 3X + 2$, can X and Y be independent?

Exercise 7.2. According to the inside of the cap on a bottle of Snaple's Mango Madness, "the number of times a cricket chirps in 15 seconds plus

37 will give you the current air temperature." How many times would you expect to hear a cricket chirp in 15 seconds when the temperature is 39°F? 124°F?

Exercise 7.3. If we constantly observe large values of one variable, call it Y, whenever we observe large values of another variable, call it X, does this mean X is part of the mechanism responsible for increases in the value of Y? If not, what are the other possibilities? To illustrate the several possibilities, give at least three real-world examples in which this statement would be false. (You'll do better at this exercise if you work on it with one or two others.)

7.2.1. Linear Regression

Equation 7.1 is an example of linear regression. The general form of linear regression is

$$Y = \mu + \beta f[X] + Z \qquad (7.2)$$

where Y is known as the *dependent* or *response variable*, X is known as the *independent variable* or *predictor*, $f[X]$ is a function of known form, μ and β are unknown *parameters*, and Z is a random variable whose expected value is zero. If it weren't for this last random component Z, then if we knew the parameters μ and β, we could plot the values of the dependent variable Y and the function $f[X]$ as a straight line on a graph. Hence the name, *linear regression*.

For the past year, the price of homes in my neighborhood could be represented as a straight line on a graph relating house prices to time: $P = \mu + \beta t$, where μ was the price of the house on the first of the year and t is the day of the year. Of course, as far as the price of any individual house was concerned, there was a lot of fluctuation around this line depending on how good a salesperson the realtor was and how desperate the owner was to sell.

If the price of my house ever reaches $700,000, I might just sell and move to Australia. Of course, a straight line might not be realistic. Prices have a way of coming down as well as going up. A better prediction formula might be $P = \mu + \beta t - \gamma t^2$, in which prices continue to rise until $\beta - \gamma t = 0$, after which they start to drop. If I knew what β and γ were or could at least get some good estimates of their value, then I could sell my house at the top of the market!

The trick is to look at a graph such as Figure 7.1 and somehow extract that information.

Note that $P = \mu + \beta t - \gamma t^2$ is another example of *linear regression*, only with three parameters rather than two. So is the formula $W = \mu + \beta H + \gamma A + Z$, where W denotes the weight of a child, H is its height, A its age, and Z, as always, is a purely random component. $W = \mu + \beta H + \gamma A + \delta AH + Z$ is still another example. The parameters μ, β, γ, and so forth are sometimes referred to as the *coefficients* of the model.

What then is a *nonlinear* regression? Here are two examples:

$$Y = \beta \log(\gamma X), \quad \text{which is linear in } \beta \text{ but nonlinear in } \gamma$$

and

$$T = \beta \cos(t + \gamma), \quad \text{which also is linear in } \beta \text{ but nonlinear in } \gamma$$

Regression models that are nonlinear in their parameters are beyond the scope of this text. The important lesson to be learned from their existence is that we need to have some idea of the functional relationship between a response variable and its predictors before we start to fit a linear regression model.

Exercise 7.4. Use R to generate a plot of the function $P = 100 + 10t - 1.5t^2$ for values of $t = 0, 1, \ldots, 10$. Does the curve reach a maximum and then turn over?

7.3. FITTING A REGRESSION EQUATION

Suppose we have determined that the *response variable Y* whose value we wish to predict is related to the value of a *predictor variable X* by the equation $E(Y) = a + bX$, and on the basis of a sample of n paired observations $(x_1, y_1), (x_2, y_2), \ldots, (x_n, y_n)$ we wish to estimate the unknown coefficients a and b. Three methods of estimation are in common use: ordinary least squares, errors-in-variables or Deming regression, and least absolute deviation. We will study all three in the next few sections.

7.3.1. Ordinary Least Squares

The ordinary least squares (OLS) technique of estimation is the most commonly used, primarily for historical reasons as its computations can be done (with some effort) by hand or with a primitive calculator. The objective of the method is to determine the parameter values that will minimize the sum of squares $\Sigma[y_i - E(Y)]^2$, where $E(Y)$ is modeled by the right-hand side of our regression equation.

In our example, $E(Y) = a + bx_i$, and so we want to find the values of a and b that will minimize $\Sigma(y_i - a - bx_i)^2$. We can readily obtain the desired estimates with the aid of the R function **lsfit()** as in the following program in which we regress systolic blood pressure (SBP) as a function of age (Age):

```
➢ Age=c(39, 47, 45,47,65,46,67,42,67,56,64,56,59,34,42)
➢ SBP=c(144,220,138,145,162,142,170,124,158,154,162,150,
  140,110,128)
➢ lsfit(Age, SBP)
$coefficients
Intercept       X
95.612512 1.047439
```

which we interpret as $\hat{E} = \hat{a} + \hat{b}\text{Age} = 95.6 + 1.04\text{Age}$.

Notice that when we report our results, we drop decimal places that convey a false impression of precision.

How good a fit is this regression line to our data? Part of our output includes a printout of the *residuals*, that is, of the differences between the values our equation would predict and what individual SBPs were actually observed.

```
$residuals
[1]    7.5373843  75.1578758  -4.7472470   0.1578758 -1.6960182
       -1.7946856
[7]    4.2091047 -15.6049314  -7.7908953  -0.2690712
       -0.6485796  -4.2690712
[13] -17.4113868 -21.2254229 -11.6049314
```

Consider the fourth residual in the series, 0.15. This is the difference between what was observed, SBP = 145, and what the regression line estimates as the expected SBP for a 47-year-old individual $E(\text{SBP}) = 95.6 + 1.04 * 47 = 144.8$. Note that we were furthest off the mark in our predictions, that is, we had the largest residual, in the case of the youngest member of our sample, a 34 year old. Oops, I made a mistake; there is a still larger residual.

Mistakes are easy to make when glancing over a series of numbers with many decimal places. You won't make a similar mistake if you first graph your results using the following R code:

```
➢ Pred = lm(SBP ~ Age)
➢ plot (Age, SBP)
➢ lines (Age,fitted(Pred))
```

Now the systolic blood pressure value of 220 for the 47 year old stands out from the rest.

The linear model R function `lm()` lends itself to the analysis of more complex regression functions. In this example and the one following, we use functions as predictors. For example, if we had observations on body temperature taken at various times of the day, we would write[1]

```
➢ lm (temp ~ I(cos (2*π*(time + 300)/1440)))
```

when we wanted to obtain a regression formula. Or, if we had data on weight and height as well as age and systolic blood pressure, we would write

```
➢ lm(SBP ~ Age+I(100*(weight/(height*height))))
```

If our regression model is $E(Y) = \mu + \alpha X + \beta W$, that is, if we have two *predictor variable*s X and W, we would write `lm(Y~X+W)`. If our regression model is $E(Y) = \mu + \alpha f[X] + \beta g[W]$, we would write `lm(Y~I(f[X])+I(g[W]))`.

Note that the `lm()` function automatically provides an estimate of the *intercept* μ. The term intercept stands for "intercept with the Y axis," since when $f[X]$ and $g[W]$ are both zero, the expected value of Y will be μ.

Sometimes, we know in advance that μ will be zero. Two good examples would be when we measure distance traveled or the amount of a substance produced in a chemical reaction as functions of time. We can force `lm()` to set $\mu = 0$ by writing `lm(Y~0+X)`.

Sometimes, we would like to introduce a cross product or interaction term into the model, $EY = \mu + \alpha X + \beta W + \gamma XW$, in which case, we would write `lm(Y~X*W)`. Note that when used in expressing a linear model formula, the arithmetic operators in R have interpretations different from their usual ones. Thus, in `Y~X*W` the "*" operator defines the regression model as $EY = \mu + \alpha X + \beta W + \gamma XW$ and not as $EY = \mu + \gamma XW$ as you might expect. If we wanted the latter model we would have to write `Y~I(X*W)`. The `I( )` function restores the usual meanings to the arithmetic operators.

---

[1] To ensure that the arithmetic operators have their accustomed interpretations—rather than those associated with the formula( ) function—it is necessary to show those functions as arguments to the **I( )** function.

Exercise 7.5. Do U.S. residents do their best to spend what they earn? Fit a regression line using OLS to the data in the accompanying table relating disposable income to expenditures in the United States from 1960 to 1982.

Economic report of the President, 1988

Year	Income 1982 ($)	Expenditures 1982 ($)
1960	6036	5561
1962	6271	5729
1964	6727	6099
1966	7280	6607
1968	7728	7003
1970	8134	7275
1972	8562	7726
1974	8867	7826
1976	9175	8272
1978	9735	8808
1980	9722	8783
1982	9725	8818

Exercise 7.6. Suppose we've measured the dry weights of chicken embryos at various intervals at gestation and recorded our findings in the following table:

Age (days)	6	7	8	9	10	11	12	13	14	15	16
Weight (g)	0.029	0.052	0.079	0.125	0.181	0.261	0.425	0.738	1.130	1.882	2.812

Create a scatter plot for the data in the table and overlay this with a plot of the regression line of weight with respect to age. Recall from Section 6.1 that the preferable way to analyze growth data is by using the logarithms of the exponentially increasing values. Overlay your scatter plot with a plot of the new regression line of log(weight) as a function of age. Which line (or model) appears to provide the better fit to the data?

Exercise 7.7. Obtain and plot the OLS regression of systolic blood pressure with respect to age after discarding the outlying value of 220 recorded for a 47-year-old individual.

Exercise 7.8. In a further study of systolic blood pressure as a function of age, the height and weight of each individual were recorded. The latter were converted to a Quetlet index using the formula QUI = 100*weight/(height)². Fit a multivariate regression line of systolic blood pressure with respect to age and the Quetlet index using the following information:

$$Age = c(41, 43, 45, 48, 49, 52, 54, 56, 57, 59, 62, 63, 65)$$

$$SBP = c(122, 120, 135, 132, 130, 148, 146, 138, 135, 166, 152, 170, 164)$$

$$QUI = c(3.25, 2.79, 2.88, 3.02, 3.10, 3.77, 2.98, 3.67, 3.17, 3.88, 3.96, 4.13, 4.01)$$

Types of Data

The linear regression model is a quantitative one. When we write $Y = 3 + 2X$, we imply that the product $2X$ will be meaningful. This will be the case if X is a metric variable. In many surveys, respondents use a nine-point *Likert scale*, where a value of "1" means they definitely disagree with a statement, and "9" means they definitely agree. Although such data are ordinal and not metric, the regression equation is still meaningful.

When one or more predictor variables are categorical, we must use a different approach. R makes it easy to accommodate categorical predictors. We first declare the categorical variable to be a factor as in Section 1.6. The regression model fit by the `lm()` function will include a different additive component for each level of the factor or categorical variable. Thus, we can include gender or race as predictors in a regression model.

Exercise 7.9. Using the Milazzo data of Exercise 6.21, express the fecal coliform level as a linear function of oxygen level and month. Solve for the coefficients using OLS.

7.3.2. Least Absolute Deviation Regression

Least absolute deviation (LAD) regression attempts to correct one of the major flaws of OLS, that of giving sometimes excessive weight to extreme values. The LAD method solves for those values of the coefficients in the regression equation for which the sum of the absolute deviations $\Sigma|y_i - R[x_i]|$ is a minimum. You'll need to install a new package in R in order to do the calculations.

The easiest way to do an installation is get connected to the internet and then type

```
➢  install.packages("quantreg")
```

The installation, which includes downloading, unzipping, and integrating the new routines, is then done automatically. The installation needs to be done once and once only. But each time before you can use the LAD routine, you'll need to load the supporting functions into computer memory by typing

```
➤ library(quantreg)
```

To perform the LAD regression, type

```
➤ rq(formula = SBP ~ Age)
```

to obtain the following output:

```
Coefficients:
(Intercept)      Age
 81.157895   1.263158
```

To display a graph, type

```
➤ f = coef(rq(SBP ~ Age))
➤ pred = f[1] + f[2]*Age
➤ plot (Age, SBP)
➤ lines (Age, pred)
```

Exercise 7.10. Create a single graph on which both the OLS and LAD regressions for the Age and SBP are displayed.

Exercise 7.11. Repeat the previous exercise after first removing the outlying value of 220 recorded for a 47-year-old individual.

Exercise 7.12. Use the LAD method using the data of Exercise 7.8 to obtain a prediction equation for systolic blood pressure as a function of age and the Quetlet index.

Exercise 7.13. The State of Washington uses an audit recovery formula in which the percentage to be recovered (the overpayment) is expressed as a linear function of the amount of the claim. The slope of this line is close to zero. Sometimes, depending on the audit sample, the slope is positive, and sometimes it is negative. Can you explain why?

7.3.3. Errors-in-Variables Regression

The need for errors-in-variables (EIV) or Deming regression is best illustrated by the struggles of a small medical device firm to bring its product to market. The first challenge was to convince regulators that its long-lasting device provided equivalent results to those of a less-efficient device already on the market. In other words, the firm needed to show that the values V recorded by its device bore a linear relation to the values W recorded by its competitor, that is, that $E(V) = a + bW$.

In contrast to the examples of regression we looked at earlier, the errors inherent in measuring W (the so-called predictor) were as large if not larger than the variation inherent in the output V of the new device.

The EIV regression method used to demonstrate equivalence differs in two respects from that of OLS:

1. With OLS, we are trying to minimize the sum of squares $\Sigma(y_{oi} - y_{pi})^2$, where y_{oi} is the ith observed value of Y and y_{pi} is the ith predicted value. With EIV, we are trying to minimize the sums of squares of errors, going both ways: $\Sigma(y_{oi} - y_{pi})^2/\text{Var } Y + \Sigma(x_{oi} - x_{pi})^2/\text{Var } X$.
2. The coefficients of the EIV regression line depend on $\lambda = \text{Var } X/\text{Var } Y$.

To compute the EIV regression coefficients, you'll need the following R function along with an estimate of the ratio λ of the variances of the two sets of measurements:

```
➢ eiv = function (X,Y, lambda){
➢ My = mean(Y)
➢ Mx = mean (X)
➢ Sxx = var(X)
➢ Syy = var(Y)
➢ Sxy = cov(X,Y)
➢ diff = lambda*Syy-Sxx
➢ root = sqrt(diff*diff+4*lambda*Sxy*Sxy)
➢ b = (diff+ root)/(2*lambda*Sxy)
➢ a = My - b*Mx
➢ list (a=a, b=b)
➢ }
```

Exercise 7.14. Are the following two sets of measurements comparable, that is, does the slope of the EIV regression line differ significantly from unity? [*Hint*: Use the sample variances to estimate lambda.]

$X = $ c(2.521, 3.341, 4.388, 5.252, 6.422, 7.443, 8.285, 9.253, 10.621, 10.405, 11.874, 13.444, 13.343, 16.402, 19.108, 19.25, 20.917, 23.409, 5.583, 5.063, 6.272, 7.469, 10.176, 6.581, 7.63)

$Y = $ c(2.362, 3.548, 4.528, 4.923, 6.443, 6.494, 8.275, 9.623, 9.646, 11.542, 10.251, 11.866, 13.388, 17.666, 17.379, 21.089, 21.296, 23.983, 5.42, 6.369, 7.899, 8.619, 11.247, 7.526, 7.653)

Exercise 7.15. Which method—OLS, LAD, or EIV—should be used to regress U as a function of W in the following cases?

(a) Some of the U values are suspect.

(b) It's not clear whether U or W is the true independent variable or whether both depend on the value of a third hidden variable.

(c) Minor errors in your predictions aren't important; large ones could be serious.

7.3.4. Assumptions

To use any of the preceding linear regression methods, the following as-yet-unstated assumptions must be satisfied:

1. *Independent random components.* In the model $y_i = \mu + \beta x_i + z_i$, the random fluctuations z_i must be independent of one another. If the z_i are not, it may be that a third variable W is influencing their values. In such a case, we would be advised to try the model $y_i = \mu + \beta x_i + \gamma w_i + \varepsilon_i$.

   When observations are made one after the other in time, it often is the case that successive errors are dependent on one another, but we can remove this dependence if we work with the increments $y_2 - y_1$, $y_3 - y_2$, and so forth. Before we start fitting our model, we would convert to these differences using the following R code:

```
➤ xdiff = c(1:length(x)-1)
➤ ydiff = c(1:length(x)-1)
➤ for (i in 1:length(x)-1){
➤ + xdiff[i]= x[i+1]-x[i]
➤ + ydiff[i]= y[i+1]-y[i]
➤ }
```

2. *Identically distributed random components.* Often, it is the case that large random fluctuations are associated with larger values of the observations. In such cases, techniques available in more advanced textbooks provide for weighting the observations when estimating model coefficients so that the least variable observations receive the highest weight.

3. *Random fluctuations follow a specific distribution.* Most statistics packages provide tests of the hypotheses that the model parameters are significantly different from zero. For example, if we enter the following instructions in R,

```
➤ Age=c(39,47,45,47,65,46,67,42,67,56,64,56,59,34,42)
➤ SBP=c(144,220,138,145,162,142,170,124,158,
  154,162,150,140,110,128)
➤ summary(lm(SBP~Age))
```

a lengthy output is displayed which includes the statements

```
Coefficients:
            Estimate Std.   Error    t value   Pr(>|t|)
(Intercept)  95.6125   29.8936      3.198    0.00699  **
Age           1.0474    0.5662      1.850    0.08717
```

This tells us that the intercept of the OLS regression is significantly different from 0 at the 1% level, but the coefficient for Age is not significant at the 5% level. The *p*-value 0.08717 is based on the assumption that all the random fluctuations come from a normal distribution. If they do not, then this *p*-value may be quite misleading as to whether or not the regression coefficient of systolic blood pressure with respect to age is significantly different from 0.

An alternate approach is to obtain a confidence interval for the coefficient by taking a series of bootstrap samples from the collection of pairs of observations (39, 144), (47, 220), . . . , (42, 128).

Exercise 7.16. Obtain a confidence interval for the regression coefficient of systolic blood pressure with respect to age using the bias-corrected-and-accelerated bootstrap described in Section 4.2.4.

7.4. PROBLEMS WITH REGRESSION

At first glance, regression seems to be a panacea for all modeling concerns. But it has a number of major limitations, just a few of which we will discuss in this section.

- The model that best fits the data we have in hand may not provide the best fit to the data we gather in the future.

• More than one linear regression model may provide a statistically significant fit to our data.

7.4.1. Goodness of Fit Versus Prediction

Two assumptions we make whenever we use a regression equation to make predictions are:

1. Relationships among the variables and, thus, the true regression line remain unchanged over time.
2. The sources of variation are the same as when we first estimated the coefficients.

We are seldom on safe ground when we attempt to use our model outside the range of predictor values for which it was developed originally. For one thing, literally every phenomenon seems to have nonlinear behavior for very small and very large values of the predictors. Treat every predicted value outside the original data range as suspect.

In my lifetime, regression curves failed to predict a drop in the sales of 78 records as a result of increasing sales of 45s, a drop in the sales of 45s as a result of increasing sales of 8-track tapes, a drop in the sales of 8-track tapes as a result of increasing sales of cassettes, nor the drop in the sales of cassettes as a result of increasing sales of CDs. It is always advisable to revalidate any model that has been in use for an extended period (see Section 7.4.2).

Finally, let us not forgot that our models are based on samples, and that sampling error must always be inherent in the modeling process.

Exercise 7.17. Redo Exercise 7.2.

7.4.2. Which Model?

The exact nature of the formula connecting two variables cannot be determined by statistical methods alone. If a linear relationship exists between two variables X and Y, then a linear relationship also exists between Y and any monotone (nondecreasing or nonincreasing) function of X. Assume X can only take positive values. If we can fit Model I: $Y = \alpha + \beta X + \varepsilon$ to the data, we also can fit Model II: $Y = \alpha' + \beta' \log[X] + \varepsilon$, and Model III: $Y = \alpha'' + \beta'' X + \gamma X^2 + \varepsilon$. It can be very difficult to determine which model if any is the "correct" one.

Five principles should guide you in choosing among models.

1. *Prevention.* The data you collect should span the entire range of interest. For example, when employing EIV regression to compare two methods of measuring glucose, it is essential to observe many pairs of observed abnormal values (characteristic of a disease process) along with the more readily available pairs of normal values. Don't allow your model to be influenced by one or two extreme values—whenever this is a possibility, employ LAD regression rather than OLS. Strive to obtain response observations at intervals throughout the relevant range of the predictor. Only when we have observations spanning the range of interest can we begin to evaluate competitive models.

2. *Think why rather than what.* In Exercise 7.6, we let our knowledge of the underlying growth process dictate the use of $\log(X)$ rather than X. As a second example, consider that had we wanted to find a relationship between the volume V and temperature T of a gas, any of the preceding three models might have been used to fit the relationship. But only one, the model $V = a + KT$, is consistent with kinetic molecular theory.

3. *Plot the residuals,* that is, plot the error or difference between the values predicted by the model and the values that were actually observed. If a pattern emerges from the plot, then modify the model to correct for the pattern. The next two exercises illustrate this approach.

Exercise 7.18. Apply Model III to the blood pressure data at the beginning of Section 7.3.1. Use R to plot the residuals with the code

```
➤  temp=lm(SBP~Age+Age*Age)
➤  plot(Age,resid(temp))
```

What does this plot suggest about the use of Model III in this context versus the simpler model that was used originally?

Exercise 7.19. Plot the residuals for the models and data of Exercise 7.6. What do you observe?

The final two guidelines are contradictory in nature:

1. *The more parameters the better the fit.* Thus, Model III is to be preferred to the two simpler models.

2. *The simpler, more straightforward model* is more likely to be correct when we come to apply it to data other than the observations in hand; thus, Models I and II are to be preferred to Model III.

Measures of Predictive Success

In order to select among models, we need to have some estimate of how successful each model will be for prediction purposes. One measure of goodness-of-fit of the model is $SSE = \Sigma(y_i - y_i^*)^2$, where y_i and y_i^* denote the ith observed value and the corresponding value obtained from the model. The smaller this sum of squares, the better the fit.

If the observations are independent, then

$$\sum\left(y_i - y_i^*\right)^2 = \sum(y_i - \bar{y})^2 - \sum\left(\bar{y}_i - y_i^*\right)^2$$

The first sum on the right-hand side of the equation is the total sum of squares (SST). Most statistics software uses as a measure of fit $R^2 = 1 - SSE/SST$. The closer the value of R^2 is to 1, the better.

The automated entry of predictors into the regression equation using R^2 runs the risk of over fitting, as R^2 is guaranteed to increase with each predictor entering the model. To compensate, one may use the adjusted R^2

$$1 - [((n - i)(1 - R^2))/(n - p)]$$

where n is the number of observations used in fitting the model, p is the number of regression coefficients in the model, and i is an indicator variable that is 1 if the model includes an intercept and 0 otherwise.

When you use the R function `summarize(lm())` it calculates values of both R-Squared (R^2) and Adjusted R-Squared.

7.4.3. Multivariable Regression

We've already studied several examples in which we utilized multiple predictors in order to obtain improved models. Sometimes, as noted in Section 7.3.4, dependence among the random errors (as seen from a plot of the residuals) may force us to use additional variables. This result is in line with our discussion of experimental design in Chapter 5. We either must control all sources of variation, must measure them, or must tolerate the "noise."

But adding more variables to the model equation creates its own set of problems. Do predictors U and V influence Y in a strictly additive fashion so that we may write $Y = \mu + \alpha U + \beta V + Z$ or, equivalently, `lm(Y~U + V)`. What if U represents the amount of fertilizer, V the total hours of sunlight, and Y is the crop yield? If there are too many cloudy days then adding fertilizer won't help a bit. The effects of fertilizer and sunlight are superadditive (or *synergistic*). A better model would be $Y = \mu + \alpha U + \beta V + \gamma UV + Z$ or `lm(Y~U*V)`.

To achieve predictive success, our observations should span the range over which we wish to make predictions. With only a single predictor, we might make just ten observations spread across the predictor's range. With two synergistic or antagonist predictors we are forced to make 10×10 observations, with three, $10 \times 10 \times 10 = 1000$ observations, and so forth. We can cheat, scattering our observations at random or in some optimal systematic fashion across the grid of possibilities, but there will always be a doubt as to our model's behavior in the unexplored areas.

The vast majority of predictors are interdependent. Changes in the value of one will be accompanied by changes in the other. (Notice that we do *not* write "changes in the value of one will cause or result in changes in the other." There may be yet a third, hidden variable responsible for all the changes.) What this means is that more than one set of coefficients may provide an equally good fit to our data. And more than one set of predictors!

As the following exercise illustrates, whether or not a given predictor will be found to make a statistically significant contribution to a model will depend on what other predictors are present.

Exercise 7.20. In order to optimize an advertising campaign for a new model of automobile by directing the campaign toward the best potential customers, a study of consumers' attitudes, interests, and opinions was commissioned. The questionnaire consisted of a number of statements covering a variety of dimensions, including consumers' attitudes toward risk, foreign-made products, product styling, spending habits, emissions, pollution, self-image, and family. The final question concerned the potential customer's attitude toward purchasing the product itself. All responses were tabulated on a nine-point Likert scale. Utilize the data below to construct a series of models as follows:

Express Purchase as a function of Fashion and Gamble.

Express Purchase as a function of Fashion, Gamble, and Ozone.

Express Purchase as a function of Fashion, Gamble, Ozone, and Pollution.

In each instance, use the summary function along with the linear-model function, as in `summary(lm())`, to determine the values of the coefficients, the associated p-values, and the values of Multiple R-Squared and Adjusted R-Squared. (As noted in the preface, for your convenience the following datasets may be downloaded from `ftp://ftp.wiley.com/public/sci_tech_med/statistics_resampling`)

Purchase = c(6, 9, 8, 3, 5, 1, 3, 3, 7, 4, 2, 8, 6, 1, 3, 6, 1, 9, 9, 7, 9, 2, 2, 8, 8, 5, 1,
3, 7, 9, 3, 6, 9, 8, 5, 4, 8, 9, 6, 2, 8, 5, 6, 5, 5, 3, 7, 6, 4, 5, 9, 2, 8, 2, 8,
7, 9, 4, 3, 3, 4, 1, 3, 6, 6, 5, 2, 4, 2, 8, 7, 7, 6, 1, 1, 9, 4, 4, 6, 9, 1, 6, 9,
6, 2, 8, 6, 3, 5, 3, 6, 8, 2, 5, 6, 7, 7, 5, 7, 6, 3, 5, 8, 8, 1, 9, 8, 8, 7, 5, 2,
2, 3, 8, 2, 2, 8, 9, 5, 6, 7, 4, 6, 5, 8, 4, 7, 8, 2, 1, 7, 9, 7, 5, 5, 9, 9, 9, 7,
3, 8, 9, 8, 4, 8, 5, 5, 8, 4, 3, 7, 1, 2, 1, 1, 7, 5, 5, 1, 4, 1, 2, 9, 7, 6, 9, 9,
6, 5, 4, 3, 6, 6, 4, 5, 7, 2, 6, 5, 6, 3, 8, 2, 5, 3, 4, 2, 3, 8, 3, 9, 1, 3, 1, 2,
5, 1, 5, 6, 7, 1, 1, 1, 4, 4, 8, 4, 7, 4, 4, 2, 6, 6, 6, 7, 2, 9, 4, 1, 9, 3, 5, 7,
2, 2, 8, 9, 2, 4, 1, 7, 1, 3, 6, 2, 6, 2, 8, 4, 4, 1, 1, 2, 2, 8, 3, 3, 3, 1, 1, 6,
8, 3, 7, 5, 9, 8, 3, 5, 6, 3, 4, 6, 1, 1, 5, 6, 6, 9, 6, 9, 9, 6, 7, 3, 8, 4, 2, 6,
4, 8, 3, 3, 6, 4, 4, 9, 5, 6, 4, 5, 3, 3, 2, 5, 9, 5, 1, 3, 4, 3, 6, 8, 1, 5, 3, 4,
8, 2, 5, 3, 2, 3, 2, 5, 8, 3, 1, 6, 3, 7, 8, 9, 2, 3, 5, 7, 7, 3, 7, 3, 9, 2, 9, 3,
9, 2, 8, 9, 5, 1, 9, 9, 1, 8, 7, 1, 4, 9, 3, 4, 9, 1, 3, 9, 1, 5, 2, 7, 9, 6, 5, 7,
4, 6, 1, 4, 2, 7, 5, 4, 5, 9, 5, 5, 5, 2, 4, 1, 8, 7, 9, 6, 8, 1, 5, 9, 9, 9, 9, 1,
3, 3, 7, 2, 5, 6, 1, 5, 8)

Fashion = c(5, 6, 8, 2, 5, 2, 5, 1, 7, 3, 5, 5, 4, 3, 3, 5, 6, 3, 4, 3, 4, 6, 4, 6, 3, 6, 5,
4, 6, 5, 5, 3, 4, 4, 4, 3, 6, 2, 3, 4, 4, 4, 5, 2, 3, 4, 5, 5, 6, 4, 5, 5, 6, 3, 4,
4, 5, 8, 4, 5, 6, 4, 2, 5, 3, 6, 2, 3, 2, 5, 3, 5, 4, 4, 5, 4, 6, 6, 5, 8, 2, 6, 5,
6, 4, 7, 4, 5, 5, 3, 6, 6, 4, 5, 5, 4, 4, 4, 4, 3, 5, 3, 3, 5, 4, 4, 5, 7, 6, 6, 4,
4, 5, 5, 2, 2, 7, 5, 1, 6, 5, 4, 7, 7, 6, 5, 6, 3, 2, 4, 5, 3, 9, 4, 4, 6, 6, 6, 9,
4, 4, 3, 3, 3, 2, 4, 4, 5, 4, 6, 6, 3, 3, 3, 5, 4, 4, 5, 4, 6, 3, 4, 6, 3, 4, 6, 4,
5, 4, 3, 3, 6, 4, 3, 3, 4, 3, 1, 4, 5, 5, 6, 2, 6, 6, 5, 5, 3, 9, 3, 3, 1, 1, 4, 3,
3, 3, 7, 6, 6, 4, 4, 1, 3, 5, 5, 4, 6, 4, 5, 5, 4, 6, 5, 6, 2, 4, 4, 3, 8, 5, 3, 6,
5, 3, 5, 3, 3, 5, 3, 2, 2, 3, 5, 5, 5, 1, 6, 5, 1, 5, 4, 4, 3, 6, 4, 4, 5, 5, 4, 5,
5, 3, 7, 4, 7, 6, 1, 5, 4, 4, 4, 3, 3, 5, 4, 7, 4, 6, 7, 6, 4, 6, 3, 4, 4, 2, 6, 3,
6, 5, 2, 2, 5, 3, 4, 4, 4, 3, 2, 4, 6, 4, 6, 5, 6, 2, 4, 2, 3, 6, 2, 6, 5, 6, 4, 4,
4, 6, 5, 5, 1, 4, 5, 5, 4, 4, 2, 3, 6, 5, 5, 2, 2, 5, 2, 5, 4, 3, 8, 3, 6, 3, 4, 3,
6, 4, 3, 4, 2, 5, 6, 4, 5, 5, 6, 4, 6, 5, 4, 3, 8, 2, 5, 5, 3, 2, 3, 5, 4, 3, 4, 3,
5, 2, 3, 1, 4, 4, 6, 6, 6, 6, 6, 6, 4, 4, 3, 4, 4, 3, 3, 5, 4, 4, 5, 4, 6, 8, 3, 3,
5, 4, 5, 4, 5, 4, 4, 6, 6)

Gamble = c(5, 4, 7, 4, 5, 4, 3, 3, 3, 6, 2, 6, 5, 4, 5, 5, 2, 7, 6, 6, 6, 4, 2, 8, 4, 4, 3,
3, 4, 5, 4, 3, 4, 6, 5, 4, 8, 9, 7, 3, 6, 4, 6, 6, 5, 3, 4, 6, 5, 4, 5, 3, 7, 3, 8,
5, 7, 5, 3, 4, 7, 4, 4, 5, 4, 6, 1, 4, 4, 9, 5, 4, 6, 4, 4, 5, 5, 5, 6, 6, 4, 4, 8,
7, 4, 5, 3, 3, 5, 3, 4, 5, 3, 5, 6, 6, 6, 5, 7, 4, 2, 3, 7, 6, 6, 4, 8, 4, 6, 3, 4,
4, 5, 8, 3, 3, 4, 5, 5, 5, 4, 5, 1, 6, 8, 5, 6, 4, 4, 6, 5, 7, 6, 5, 6, 7, 7, 6, 6,
4, 7, 6, 6, 5, 7, 6, 6, 5, 2, 5, 5, 4, 3, 3, 4, 6, 4, 4, 4, 3, 5, 2, 6, 4, 4, 6, 7,
6, 5, 4, 4, 7, 4, 7, 8, 5, 4, 5, 5, 3, 2, 4, 3, 6, 2, 3, 6, 5, 2, 4, 6, 2, 2, 1, 4,
3, 5, 5, 4, 6, 2, 3, 3, 5, 3, 6, 5, 6, 5, 3, 4, 6, 5, 5, 5, 3, 7, 7, 2, 6, 4, 2, 7,
2, 6, 3, 6, 2, 5, 3, 7, 3, 4, 2, 3, 7, 3, 6, 3, 7, 2, 4, 4, 4, 8, 3, 4, 4, 3, 1, 4,
7, 5, 4, 5, 8, 4, 6, 4, 6, 4, 4, 5, 4, 2, 6, 5, 5, 7, 2, 7, 4, 5, 6, 5, 3, 3, 2, 5,
3, 6, 1, 5, 6, 6, 5, 8, 6, 6, 5, 6, 4, 4, 6, 4, 7, 4, 4, 5, 4, 3, 7, 8, 1, 4, 4, 7,
4, 5, 4, 5, 1, 3, 4, 4, 4, 5, 3, 5, 5, 4, 7, 6, 3, 6, 4, 6, 5, 3, 4, 5, 7, 4, 5, 4,

5, 3, 7, 6, 4, 6, 4, 8, 3, 4, 2, 5, 5, 5, 4, 5, 6, 3, 5, 8, 4, 5, 2, 5, 4, 5, 6, 3,
3, 3, 1, 5, 3, 4, 7, 4, 4, 6, 4, 3, 5, 3, 4, 4, 8, 6, 7, 4, 6, 4, 5, 4, 6, 8, 7, 2,
5, 4, 7, 4, 5, 6, 4, 6, 6)

Ozone = c(4, 7, 7, 3, 4, 2, 3, 4, 2, 5, 4, 6, 2, 3, 6, 3, 7, 8, 7, 5, 5, 5, 5, 8, 5, 5, 5, 1,
4, 6, 6, 9, 2, 6, 3, 4, 6, 6, 7, 5, 6, 6, 6, 6, 4, 3, 5, 6, 5, 5, 3, 5, 4, 5, 6, 8, 3, 8,
5, 6, 4, 4, 4, 3, 7, 8, 5, 3, 6, 6, 8, 6, 4, 5, 6, 4, 3, 6, 6, 3, 5, 5, 6, 7, 6, 6, 7,
3, 4, 5, 5, 6, 5, 3, 3, 5, 6, 3, 4, 6, 5, 5, 6, 6, 5, 9, 8, 5, 5, 4, 8, 4, 3, 5, 5, 4,
6, 8, 8, 4, 7, 5, 9, 2, 2, 5, 2, 7, 7, 2, 4, 4, 6, 3, 7, 7, 4, 3, 6, 3, 6, 6, 7, 3, 5,
5, 4, 3, 6, 4, 5, 6, 5, 5, 4, 7, 5, 5, 2, 4, 7, 5, 5, 5, 4, 6, 5, 5, 5, 7, 5, 3, 6, 5,
6, 6, 4, 4, 2, 6, 6, 4, 8, 3, 5, 3, 3, 5, 5, 6, 5, 7, 4, 1, 3, 4, 6, 4, 3, 8, 5, 2, 7,
1, 5, 3, 7, 5, 4, 3, 7, 4, 2, 8, 7, 4, 3, 6, 7, 6, 6, 7, 9, 9, 3, 7, 6, 6, 4, 5, 6, 6,
4, 6, 5, 7, 5, 4, 6, 5, 6, 5, 5, 5, 4, 4, 6, 9, 3, 3, 2, 5, 5, 5, 7, 3, 6, 4, 5, 7, 5,
4, 5, 5, 6, 6, 7, 4, 4, 4, 4, 2, 7, 4, 5, 4, 4, 5, 3, 6, 4, 7, 6, 4, 6, 5, 4, 5, 5, 4,
5, 7, 1, 3, 8, 6, 7, 5, 5, 5, 4, 5, 6, 5, 3, 5, 2, 3, 3, 4, 3, 3, 5, 5, 7, 7, 5, 6, 6,
6, 4, 7, 5, 7, 5, 8, 7, 7, 4, 5, 6, 6, 4, 9, 8, 5, 6, 6, 4, 4, 5, 4, 6, 3, 5, 4, 5, 8,
6, 6, 5, 3, 6, 7, 4, 7, 5, 4, 3, 6, 4, 6, 6, 4, 5, 5, 3, 7, 4, 6, 7, 3, 5, 6, 4, 9, 6,
3, 5, 7, 4, 5, 3, 7, 3, 3, 6, 6, 4, 6, 6, 6, 5, 5, 9, 4, 3, 6, 3, 4, 6)

Pollution = c(5, 7, 7, 4, 5, 1, 3, 6, 3, 5, 5, 6, 2, 2, 7, 2, 6, 7, 7, 5, 5, 5, 6, 8, 6, 4, 5,
1, 4, 6, 6, 9, 3, 6, 3, 6, 4, 6, 8, 6, 6, 5, 6, 5, 3, 3, 8, 7, 7, 3, 4, 5, 5, 6, 8,
3, 8, 5, 6, 5, 4, 6, 5, 7, 7, 6, 4, 6, 5, 8, 5, 6, 6, 6, 4, 4, 5, 6, 5, 6, 5, 6, 8,
5, 5, 6, 5, 4, 5, 6, 5, 6, 3, 4, 6, 6, 5, 6, 6, 5, 4, 6, 8, 4, 9, 7, 6, 4, 5, 9, 4,
4, 4, 5, 4, 5, 7, 8, 3, 7, 7, 7, 2, 2, 5, 3, 5, 7, 4, 5, 5, 7, 5, 5, 5, 6, 5, 4, 7, 4,
7, 7, 7, 4, 5, 5, 5, 3, 6, 5, 5, 7, 6, 4, 6, 7, 4, 6, 4, 4, 7, 6, 6, 7, 4, 6, 5, 6,
5, 6, 5, 4, 6, 6, 5, 6, 5, 6, 3, 6, 6, 5, 7, 4, 3, 3, 4, 6, 5, 5, 6, 6, 5, 2, 4, 4,
6, 2, 3, 6, 5, 4, 6, 2, 5, 2, 8, 4, 5, 4, 7, 5, 1, 7, 5, 6, 4, 5, 7, 7, 5, 6, 8, 8,
5, 7, 5, 5, 5, 6, 6, 5, 6, 4, 7, 5, 6, 5, 4, 4, 5, 5, 6, 7, 5, 6, 4, 6, 4, 8, 4, 4, 2, 4,
5, 6, 6, 4, 5, 6, 4, 6, 6, 4, 5, 4, 7, 6, 7, 5, 5, 5, 4, 2, 6, 3, 5, 3, 4, 5, 3, 5,
4, 7, 7, 4, 6, 5, 5, 4, 6, 4, 6, 6, 1, 4, 7, 5, 8, 3, 6, 4, 4, 4, 4, 6, 7, 6, 5, 3, 3,
5, 4, 4, 4, 6, 5, 7, 6, 4, 7, 6, 6, 4, 8, 4, 8, 6, 7, 8, 8, 4, 5, 4, 6, 5, 7, 7, 5,
6, 6, 5, 6, 6, 4, 6, 6, 5, 4, 7, 6, 6, 5, 6, 3, 4, 9, 5, 6, 5, 3, 5, 6, 4, 6, 4, 4,
6, 6, 3, 7, 5, 5, 7, 4, 6, 6, 4, 9, 6, 5, 6, 7, 4, 5, 3, 5, 4, 3, 7, 6, 6, 6, 6, 5,
6, 6, 9, 6, 3, 6, 3, 5, 7)

As noted earlier, when you use the R function `summarize(lm())` it yields values of Multiple *R*-Squared and Adjusted *R*-Squared. Both are measures of how satisfactory a fit the model provides to the data. The second measure adjusts for the number of terms in the model; as this exercise reveals, the greater the number of terms, the greater the difference between the two measures.

The R function `step()` automates the variable selection procedure. Unless you specify an alternate method, `step()` will start the process by including all the variables you've listed in the model. Then it will eliminate the least significant predictor, continuing a step at a time until all the terms that remain in the model are significant at the 5% level.

Applied to the data of Exercise 7.20, one obtains the following results:

```
➤   summary(step(lm(Purchase ~ Fashion + Gamble +
      Ozone + Pollution)))
Start:  AIC= 561.29
  Purchase ~ Fashion + Gamble + Ozone + Pollution

              Df  Sum of Sq  RSS       AIC
- Ozone        1     1.82    1588.97   559.75
- Pollution    1     2.16    1589.31   559.84
<none>                       1587.15   561.29
- Fashion      1   151.99    1739.14   595.87
- Gamble       1   689.60    2276.75   703.62

Step:  AIC = 559.75
  Purchase ~ Fashion + Gamble + Pollution

              Df  Sum of Sq  RSS       AIC
<none>                       1588.97   559.75
- Pollution    1    17.08    1606.05   562.03
- Fashion      1   154.63    1743.60   594.90
- Gamble       1   706.18    2295.15   704.84

Call:
lm(formula = Purchase ~ Fashion + Gamble + Pollution)

Rsiduals:
   Min      1Q     Median     3Q       Max
 -5.5161 -1.3955  -0.1115   1.4397   5.0325

Coefficients:
             Estimate Std. Error t value Pr(>|t|)
(Intercept)  -1.54399  0.49825   -3.099  0.00208  **
Fashion       0.41960  0.06759    6.208  1.36e-09 ***
Gamble        0.85291  0.06429   13.266  < 2e-16  ***
Pollution     0.14047  0.06809    2.063  0.03977  *
--
Signif. codes:  0`***' 0.001`**' 0.01`*' 0.05`.' 0.1` ` 1

Residual standard error: 2.003 on 396 degrees of freedom
Multiple R-Squared: 0.4051,    Adjusted R-squared: 0.4005
F-statistic: 89.87 on 3 and 396 DF,   p-value: <2.2e-16
```

From which we extract the model

E(Purchase) = −1.54 + 0.42 Fashion + 0.85 Gamble + 0.14 Pollution

In terms of the original questionnaire, this means that the individuals most likely to purchase an automobile of this particular make and model are those who definitely agreed with the statements "Life is too short not to take some gambles," "When I must choose, I dress for fashion, not comfort," and "I think the government is doing too much to control pollution."

Exercise 7.21. The marketing study also included 11 more questions, the responses to which we've tabulated below, again using a nine-point Likert scale. Use the R stepwise regression function to find the best prediction model for Attitude using the answers to all 15 questions.

Today = c(4, 4, 4, 5, 4, 4, 5, 5, 6, 5, 6, 4, 2, 5, 6, 2, 5, 6, 6, 6, 7, 6, 5, 4, 6, 7, 2, 6,
2, 6, 4, 4, 5, 6, 4, 6, 5, 5, 4, 4, 2, 6, 3, 5, 5, 4, 4, 5, 6, 2, 4, 4, 4, 2, 2, 7, 4,
2, 5, 8, 7, 6, 5, 6, 6, 4, 6, 3, 6, 5, 3, 2, 7, 3, 4, 6, 4, 3, 7, 5, 5, 5, 4, 3, 4, 5,
5, 4, 3, 5, 5, 5, 6, 4, 5, 3, 7, 6, 6, 3, 5, 4, 6, 5, 5, 3, 6, 5, 9, 2, 6, 3, 6, 7, 4, 3,
1, 6, 5, 3, 6, 4, 5, 4, 6, 4, 2, 5, 1, 1, 4, 1, 2, 5, 4, 4, 4, 5, 3, 6, 6, 3, 5, 2, 4,
2, 4, 3, 6, 3, 7, 5, 4, 3, 4, 4, 5, 3, 4, 6, 9, 3, 2, 5, 5, 6, 6, 4, 7, 6, 5, 4, 7, 5,
4, 4, 4, 5, 6, 4, 4, 1, 2, 7, 7, 3, 4, 6, 6, 5, 3, 3, 5, 6, 5, 4, 4, 3, 6, 3, 3, 8, 2,
5, 4, 3, 5, 5, 2, 4, 5, 7, 4, 5, 3, 4, 3, 5, 4, 5, 6, 4, 5, 4, 4, 6, 3, 4, 5, 7, 3, 4,
4, 2, 5, 5, 6, 6, 5, 4, 6, 3, 4, 4, 2, 4, 5, 5, 5, 5, 5, 4, 3, 6, 3, 5, 1, 4, 6, 3, 6,
5, 4, 3, 4, 4, 5, 5, 6, 5, 5, 2, 5, 3, 3, 6, 8, 2, 4, 7, 4, 3, 4, 3, 3, 4, 3, 7, 4, 8,
7, 5, 2, 5, 2, 2, 7, 5, 4, 4, 4, 7, 5, 5, 3, 5, 4, 5, 6, 5, 5, 4, 7, 4, 4, 3, 6, 5, 6,
4, 5, 7, 6, 2, 6, 7, 7, 7, 1, 2, 6, 6, 3, 4, 6, 5, 4, 2, 6, 6, 6, 3, 3, 7, 2, 4, 4, 4,
4, 6, 4, 4, 6, 5, 6, 3, 3, 8, 3, 5, 5, 3, 6, 5, 4, 5, 5, 4, 3, 2, 4, 5, 1, 5, 5, 6, 5,
4, 5, 4, 3, 4, 3, 4, 6, 5, 6, 3, 6, 2, 5, 5, 6, 3, 6, 7, 5, 5, 4, 4, 4)

Coupons = c(4, 2, 3, 6, 5, 5, 6, 3, 5, 5, 6, 4, 3, 4, 6, 4, 5, 6, 4, 5, 7, 5, 4, 5, 5, 6, 1,
5, 3, 7, 5, 5, 4, 5, 3, 6, 6, 5, 3, 4, 4, 7, 4, 6, 5, 5, 3, 4, 7, 1, 4, 5, 6, 3, 3,
6, 5, 3, 6, 5, 7, 7, 4, 5, 5, 4, 6, 3, 6, 4, 3, 3, 7, 2, 5, 5, 4, 4, 8, 4, 4, 4, 5,
4, 4, 7, 5, 4, 3, 7, 5, 5, 4, 6, 6, 4, 6, 5, 6, 3, 4, 4, 7, 4, 4, 6, 6, 9, 4, 6, 2,
7, 7, 4, 4, 3, 4, 3, 3, 4, 5, 4, 4, 6, 4, 2, 4, 1, 2, 5, 2, 3, 5, 3, 3, 5, 5, 5, 5,
6, 3, 3, 3, 4, 3, 3, 4, 6, 3, 5, 4, 3, 4, 5, 4, 4, 3, 5, 5, 9, 3, 4, 6, 6, 6, 6, 4,
6, 5, 6, 3, 6, 5, 2, 5, 4, 5, 5, 4, 5, 2, 3, 5, 7, 4, 4, 6, 6, 3, 2, 3, 5, 8, 6, 5,
5, 5, 7, 5, 4, 5, 2, 4, 4, 2, 5, 3, 2, 2, 6, 7, 3, 5, 3, 4, 3, 5, 4, 5, 5, 4, 5, 4,
4, 6, 3, 3, 5, 7, 4, 6, 5, 3, 5, 5, 6, 4, 6, 3, 5, 2, 5, 3, 2, 4, 5, 5, 4, 4, 5, 4,
5, 5, 5, 4, 3, 4, 5, 5, 4, 5, 5, 4, 4, 4, 4, 6, 6, 4, 5, 2, 4, 2, 3, 6, 8, 1, 5, 6,
5, 3, 5, 3, 6, 5, 5, 6, 4, 7, 7, 6, 3, 3, 3, 3, 5, 5, 5, 5, 6, 7, 4, 5, 3, 4, 3, 3,
6, 4, 3, 5, 8, 6, 5, 4, 8, 5, 7, 3, 5, 6, 5, 1, 7, 5, 6, 5, 1, 2, 5, 5, 3, 3, 5, 5,

6, 4, 5, 5, 7, 4, 5, 5, 3, 4, 4, 3, 4, 6, 5, 3, 6, 7, 5, 4, 2, 8, 2, 6, 5, 2, 5, 6,
5, 4, 4, 5, 5, 4, 2, 4, 3, 6, 6, 7, 5, 4, 5, 3, 4, 5, 4, 4, 6, 5, 7, 4, 7, 4, 5, 5,
7, 2, 6, 7, 5, 5, 3, 5, 4)

IntRates = c(6, 1, 3, 6, 6, 2, 6, 3, 5, 5, 5, 4, 3, 3, 4, 4, 6, 6, 5, 3, 6, 6, 3, 6, 5, 6, 2,
6, 3, 6, 4, 4, 4, 5, 4, 6, 6, 6, 3, 5, 4, 7, 4, 7, 5, 7, 4, 5, 7, 2, 5, 6, 6, 4, 3,
7, 4, 3, 7, 7, 7, 6, 5, 6, 5, 4, 6, 4, 6, 4, 4, 4, 7, 3, 5, 5, 3, 4, 9, 3, 5, 3, 5,
6, 4, 7, 5, 5, 4, 5, 6, 4, 5, 5, 7, 5, 6, 5, 7, 5, 4, 5, 7, 5, 4, 7, 5, 9, 4, 7, 3,
7, 6, 5, 5, 4, 3, 4, 3, 3, 5, 3, 5, 5, 4, 3, 5, 4, 4, 4, 2, 3, 5, 3, 3, 4, 5, 6, 6,
4, 4, 4, 4, 5, 3, 4, 4, 6, 3, 5, 4, 4, 4, 5, 3, 5, 3, 6, 5, 9, 4, 5, 6, 6, 5, 6, 4,
7, 5, 7, 3, 6, 5, 2, 4, 4, 5, 5, 5, 5, 2, 3, 5, 8, 4, 3, 6, 7, 4, 4, 5, 6, 7, 6, 6,
6, 6, 6, 6, 4, 5, 2, 5, 3, 3, 6, 3, 2, 2, 7, 6, 4, 4, 3, 5, 4, 6, 2, 6, 5, 4, 7, 6,
3, 6, 4, 3, 5, 8, 5, 6, 5, 4, 6, 4, 5, 5, 7, 4, 4, 4, 7, 3, 2, 5, 5, 5, 5, 4, 5, 3,
4, 4, 5, 5, 3, 5, 6, 5, 5, 4, 7, 3, 3, 3, 4, 6, 5, 4, 6, 2, 4, 3, 3, 8, 9, 1, 7, 5,
6, 4, 4, 6, 7, 5, 5, 6, 4, 8, 7, 6, 5, 4, 4, 4, 4, 6, 5, 6, 6, 8, 5, 4, 3, 5, 3, 4,
7, 3, 2, 5, 8, 5, 5, 5, 8, 5, 8, 3, 6, 7, 5, 3, 8, 6, 5, 5, 1, 3, 4, 5, 5, 3, 4, 6,
4, 5, 4, 5, 7, 5, 5, 5, 3, 4, 5, 3, 5, 7, 7, 4, 5, 6, 4, 5, 1, 6, 3, 6, 5, 3, 5, 5,
5, 4, 5, 4, 4, 4, 2, 6, 3, 6, 5, 6, 5, 4, 5, 2, 5, 5, 4, 3, 6, 4, 8, 3, 7, 4, 5, 4,
7, 1, 7, 7, 6, 5, 4, 6, 4)

Selfconf = c(6, 7, 4, 6, 6, 6, 6, 3, 5, 6, 4, 1, 5, 5, 7, 3, 4, 6, 7, 5, 5, 4, 6, 4, 8, 5, 6,
5, 3, 5, 4, 4, 5, 8, 3, 5, 6, 3, 6, 5, 4, 7, 5, 2, 6, 5, 5, 6, 3, 5, 5, 4, 8, 6, 4,
7, 4, 5, 3, 4, 5, 3, 5, 7, 8, 6, 6, 3, 6, 6, 5, 3, 4, 5, 4, 6, 4, 6, 5, 5, 4, 5, 6,
7, 5, 6, 4, 6, 4, 3, 4, 2, 4, 1, 5, 5, 5, 6, 5, 5, 4, 4, 3, 7, 5, 5, 5, 5, 5, 4, 3,
6, 3, 5, 3, 4, 7, 6, 5, 5, 4, 3, 5, 3, 6, 5, 4, 5, 2, 5, 5, 5, 3, 6, 7, 5, 4, 6, 5,
6, 5, 3, 6, 5, 3, 6, 6, 5, 6, 6, 1, 4, 4, 9, 7, 5, 8, 7, 4, 3, 3, 6, 6, 8, 3, 7, 6,
4, 7, 4, 6, 6, 5, 6, 3, 5, 5, 5, 3, 5, 6, 6, 6, 4, 4, 5, 4, 5, 5, 3, 6, 3, 6, 4, 4,
5, 3, 6, 6, 5, 1, 4, 6, 4, 6, 4, 3, 6, 6, 5, 3, 5, 6, 6, 5, 4, 6, 5, 3, 5, 5, 5, 7,
4, 7, 3, 5, 2, 3, 2, 3, 4, 4, 5, 7, 3, 6, 6, 6, 7, 4, 3, 4, 4, 5, 2, 8, 5, 2, 5, 6,
7, 1, 4, 5, 2, 7, 6, 6, 3, 4, 2, 4, 6, 6, 3, 2, 6, 3, 4, 5, 5, 5, 4, 3, 6, 3, 4, 4,
5, 7, 7, 5, 3, 2, 6, 4, 1, 5, 4, 5, 5, 4, 5, 7, 3, 6, 6, 8, 2, 5, 6, 4, 5, 7, 3, 5,
6, 6, 4, 4, 6, 4, 4, 6, 4, 4, 4, 7, 6, 4, 6, 3, 4, 5, 3, 4, 6, 6, 4, 5, 7, 6, 6, 4,
4, 4, 4, 6, 5, 7, 4, 7, 4, 8, 6, 6, 6, 7, 4, 3, 7, 4, 3, 4, 4, 6, 6, 5, 5, 3, 3, 5,
5, 4, 4, 5, 6, 4, 8, 2, 3, 3, 2, 6, 7, 2, 7, 7, 4, 6, 6, 5, 4, 3, 3, 7, 6, 5, 5, 1,
4, 5, 8, 5, 6, 3, 5, 8, 4)

Leader = c(5, 6, 6, 6, 7, 7, 6, 5, 6, 7, 6, 3, 5, 6, 7, 4, 4, 7, 6, 6, 6, 5, 6, 5, 9, 5, 8, 6,
3, 5, 5, 6, 5, 9, 4, 4, 7, 5, 7, 6, 5, 6, 7, 3, 7, 6, 7, 7, 4, 8, 7, 6, 8, 6, 5, 7, 5,
5, 5, 5, 6, 4, 6, 8, 8, 5, 6, 6, 6, 7, 5, 4, 5, 6, 4, 6, 6, 6, 6, 6, 4, 6, 6, 8, 5, 6,
6, 5, 6, 5, 4, 3, 5, 3, 7, 5, 6, 6, 6, 6, 4, 5, 5, 7, 7, 6, 6, 6, 6, 5, 3, 8, 5, 6, 3,
5, 7, 6, 4, 6, 4, 4, 6, 4, 7, 5, 4, 7, 4, 6, 5, 5, 5, 8, 7, 6, 6, 6, 5, 6, 5, 3, 7, 4,
4, 6, 7, 5, 5, 6, 3, 5, 5, 9, 8, 8, 7, 6, 5, 3, 4, 8, 6, 9, 5, 6, 5, 5, 6, 6, 9, 5, 6,
5, 3, 7, 7, 5, 5, 6, 7, 7, 7, 5, 4, 6, 6, 6, 7, 5, 6, 4, 8, 4, 5, 6, 3, 6, 7, 5, 1, 6,

5, 6, 7, 5, 4, 6, 7, 5, 3, 5, 6, 7, 6, 5, 7, 7, 5, 6, 5, 5, 7, 5, 7, 5, 7, 2, 6, 3, 5,
5, 6, 6, 7, 3, 9, 7, 6, 8, 5, 4, 5, 4, 5, 4, 9, 6, 4, 6, 5, 8, 3, 5, 5, 4, 6, 6, 7, 4,
5, 4, 5, 7, 6, 3, 1, 7, 3, 5, 5, 6, 6, 5, 3, 7, 3, 5, 4, 5, 7, 8, 4, 3, 3, 8, 5, 2, 6,
5, 7, 5, 3, 6, 8, 3, 8, 5, 8, 2, 6, 6, 5, 6, 9, 3, 6, 7, 6, 5, 3, 5, 5, 6, 6, 5, 5, 4,
7, 7, 4, 5, 4, 5, 5, 3, 7, 6, 6, 4, 7, 7, 6, 6, 6, 6, 5, 4, 7, 5, 8, 4, 7, 6, 8, 8, 7,
6, 8, 5, 5, 7, 6, 4, 6, 6, 7, 8, 7, 5, 3, 4, 6, 7, 6, 4, 4, 7, 6, 8, 3, 4, 4, 4, 7, 9,
4, 8, 7, 4, 5, 7, 5, 5, 4, 5, 7, 8, 7, 6, 1, 4, 6, 7, 5, 7, 4, 7, 7, 6)

Trip = c(4, 5, 6, 4, 5, 2, 3, 5, 4, 2, 3, 6, 4, 3, 2, 7, 1, 6, 9, 6, 7, 4, 1, 6, 3, 3, 4, 3, 4,
5, 5, 3, 6, 4, 4, 3, 4, 6, 7, 2, 4, 4, 3, 7, 4, 3, 4, 3, 3, 5, 6, 2, 5, 1, 7, 2, 5, 4, 4,
2, 3, 3, 3, 5, 3, 4, 4, 7, 5, 6, 5, 5, 5, 2, 2, 7, 4, 2, 5, 6, 3, 4, 4, 4, 4, 6, 5, 3, 5,
3, 4, 4, 3, 3, 5, 4, 4, 2, 4, 3, 5, 5, 6, 4, 3, 4, 4, 5, 4, 6, 4, 3, 2, 5, 4, 5, 4, 6, 4,
5, 6, 5, 3, 4, 6, 4, 5, 5, 3, 3, 5, 6, 4, 2, 5, 5, 3, 6, 5, 5, 7, 3, 4, 4, 4, 4, 4, 5, 5,
4, 5, 2, 3, 2, 4, 6, 4, 4, 2, 2, 2, 5, 6, 3, 4, 7, 6, 3, 5, 5, 3, 3, 4, 3, 5, 4, 4, 4, 4,
4, 3, 7, 4, 4, 3, 3, 2, 4, 6, 4, 6, 1, 2, 1, 2, 4, 5, 5, 5, 3, 2, 1, 4, 4, 4, 3, 3, 5, 4,
7, 4, 8, 5, 4, 5, 4, 7, 6, 2, 6, 4, 6, 5, 4, 3, 6, 4, 3, 4, 4, 5, 4, 3, 4, 3, 3, 6, 6, 5,
5, 3, 4, 1, 3, 4, 2, 4, 4, 1, 2, 4, 5, 3, 4, 4, 5, 5, 4, 3, 5, 6, 5, 5, 3, 4, 3, 4, 6, 6,
6, 7, 6, 5, 3, 6, 5, 5, 5, 3, 5, 4, 3, 2, 4, 5, 3, 6, 3, 3, 4, 5, 4, 2, 2, 2, 5, 4, 4, 4,
6, 5, 7, 7, 2, 5, 2, 2, 3, 4, 5, 2, 2, 2, 1, 2, 6, 4, 2, 6, 3, 4, 5, 5, 3, 4, 5, 4, 6, 4,
1, 5, 6, 4, 2, 3, 2, 2, 6, 5, 4, 2, 7, 7, 4, 7, 7, 3, 3, 5, 5, 3, 4, 2, 4, 4, 4, 2, 3, 4,
5, 4, 4, 4, 5, 4, 5, 5, 3, 3, 5, 3, 5, 4, 3, 5, 5, 6, 5, 4, 6, 7, 5, 5, 8, 1, 4, 7, 5, 4,
7, 4, 2, 4, 2, 2, 6, 6, 1, 5, 2)

Change = c(3, 5, 5, 4, 4, 2, 5, 6, 5, 3, 3, 6, 4, 4, 3, 8, 2, 7, 9, 5, 8, 3, 2, 6, 4, 3, 4,
4, 4, 5, 6, 3, 6, 4, 4, 4, 5, 7, 7, 3, 4, 5, 4, 7, 4, 3, 4, 4, 4, 5, 6, 2, 4, 1, 6,
2, 4, 4, 4, 3, 4, 4, 4, 6, 3, 4, 4, 8, 6, 7, 5, 5, 4, 1, 3, 9, 4, 2, 4, 7, 3, 4, 4,
4, 5, 6, 6, 3, 5, 2, 4, 3, 3, 4, 5, 5, 4, 3, 5, 3, 4, 5, 6, 4, 4, 5, 5, 6, 4, 5, 4,
3, 2, 6, 5, 5, 5, 7, 5, 4, 6, 4, 4, 4, 6, 4, 5, 4, 4, 4, 6, 6, 4, 2, 5, 5, 3, 6, 6,
5, 5, 3, 4, 5, 5, 5, 5, 6, 5, 5, 6, 3, 3, 2, 4, 6, 4, 4, 3, 3, 2, 4, 6, 3, 3, 7, 7,
3, 6, 6, 3, 3, 6, 3, 5, 4, 4, 6, 4, 3, 3, 6, 3, 4, 3, 3, 2, 5, 7, 5, 6, 1, 1, 2, 2,
4, 5, 5, 5, 3, 2, 1, 4, 5, 5, 4, 4, 6, 5, 7, 4, 8, 6, 3, 5, 3, 6, 6, 3, 7, 5, 7, 5,
5, 3, 7, 3, 4, 4, 4, 4, 4, 3, 4, 3, 3, 7, 7, 5, 5, 3, 3, 1, 3, 4, 2, 4, 4, 1, 3, 4,
5, 4, 5, 5, 6, 5, 5, 4, 5, 7, 5, 5, 3, 5, 5, 6, 6, 6, 5, 8, 6, 4, 2, 6, 5, 5, 5, 3,
6, 4, 3, 3, 5, 5, 3, 6, 4, 3, 5, 5, 3, 3, 2, 2, 5, 4, 4, 5, 7, 6, 7, 7, 2, 5, 3, 3,
3, 4, 5, 2, 2, 3, 1, 1, 6, 4, 2, 6, 3, 4, 4, 6, 4, 4, 5, 3, 7, 4, 3, 5, 6, 4, 3, 2,
3, 2, 6, 7, 5, 3, 7, 7, 5, 7, 7, 5, 3, 5, 6, 4, 5, 3, 4, 5, 3, 2, 4, 4, 5, 4, 4, 5,
5, 5, 5, 6, 3, 3, 5, 3, 5, 4, 3, 5, 5, 6, 5, 3, 6, 8, 6, 5, 7, 1, 4, 7, 6, 4, 6, 3,
2, 4, 2, 3, 6, 6, 1, 6, 4)

Pioneer = c(3, 6, 5, 5, 4, 4, 7, 7, 5, 2, 6, 5, 5, 3, 3, 9, 3, 5, 7, 2, 6, 3, 3, 5, 5, 4, 5,
6, 2, 4, 6, 2, 6, 3, 6, 5, 4, 6, 5, 3, 4, 5, 5, 7, 4, 2, 4, 4, 7, 5, 6, 3, 4, 2, 6,
6, 4, 3, 4, 2, 5, 4, 6, 5, 5, 4, 4, 7, 5, 7, 6, 6, 5, 1, 4, 7, 4, 2, 7, 6, 5, 7, 5,
3, 4, 5, 5, 2, 3, 4, 6, 4, 4, 5, 5, 6, 4, 5, 6, 4, 5, 4, 7, 4, 5, 6, 4, 7, 3, 7, 4,

4, 4, 6, 5, 5, 4, 6, 8, 5, 5, 4, 3, 6, 7, 5, 4, 5, 4, 6, 6, 4, 6, 5, 6, 8, 3, 4, 6,
6, 5, 3, 5, 4, 6, 5, 4, 3, 5, 6, 6, 5, 6, 4, 4, 6, 5, 3, 3, 3, 4, 6, 4, 4, 4, 6, 6,
4, 4, 4, 5, 3, 7, 2, 3, 5, 6, 4, 4, 6, 4, 5, 4, 5, 3, 6, 3, 7, 5, 7, 7, 5, 3, 3, 3,
4, 4, 4, 4, 5, 4, 2, 5, 4, 6, 5, 6, 3, 6, 7, 4, 8, 7, 4, 6, 5, 3, 5, 3, 4, 4, 7, 6,
5, 5, 5, 3, 4, 3, 4, 5, 5, 3, 5, 3, 3, 7, 7, 6, 6, 5, 4, 1, 4, 5, 3, 5, 4, 5, 4, 5,
6, 5, 5, 5, 5, 6, 4, 6, 4, 6, 7, 7, 4, 6, 3, 5, 5, 5, 7, 6, 3, 5, 4, 5, 5, 5, 4, 5,
6, 6, 2, 4, 5, 6, 4, 6, 4, 4, 8, 4, 6, 3, 5, 2, 8, 6, 4, 4, 4, 5, 5, 6, 4, 5, 1, 3,
6, 3, 6, 3, 4, 4, 5, 3, 5, 6, 4, 6, 3, 6, 5, 6, 3, 5, 7, 4, 7, 4, 4, 5, 5, 4, 4, 4,
1, 1, 5, 6, 5, 3, 5, 6, 6, 7, 7, 5, 4, 4, 5, 7, 6, 4, 4, 7, 3, 3, 4, 3, 5, 4, 4, 5,
4, 5, 4, 6, 2, 3, 4, 4, 5, 3, 4, 5, 4, 4, 4, 4, 6, 7, 5, 6, 5, 5, 5, 5, 5, 6, 6, 5,
5, 4, 5, 4, 9, 8, 2, 7, 5)

Work = c(3, 4, 5, 5, 7, 7, 5, 4, 5, 7, 5, 1, 5, 8, 2, 7, 4, 7, 3, 6, 6, 6, 5, 5, 6, 4, 5, 5,
5, 4, 5, 3, 8, 7, 4, 4, 7, 6, 8, 4, 4, 5, 5, 6, 3, 6, 6, 5, 7, 9, 4, 5, 6, 4, 3, 6, 8,
3, 5, 8, 5, 5, 7, 5, 6, 3, 5, 6, 5, 6, 6, 5, 8, 5, 6, 5, 5, 6, 7, 5, 5, 3, 7, 5, 7, 6,
4, 6, 4, 1, 7, 3, 6, 5, 7, 5, 4, 6, 5, 5, 4, 6, 6, 6, 7, 5, 5, 6, 3, 7, 3, 7, 5, 6, 7,
6, 3, 8, 6, 5, 7, 6, 7, 7, 6, 3, 8, 5, 3, 8, 7, 6, 7, 6, 8, 9, 4, 6, 4, 7, 7, 3, 7, 6,
4, 5, 5, 4, 7, 7, 9, 6, 7, 5, 5, 6, 7, 6, 7, 7, 5, 5, 5, 4, 8, 7, 7, 6, 4, 4, 6, 5, 7,
4, 8, 4, 5, 5, 9, 4, 3, 5, 3, 6, 4, 7, 5, 6, 4, 6, 9, 4, 4, 7, 6, 7, 8, 4, 5, 5, 5, 4,
5, 5, 7, 5, 6, 4, 4, 6, 5, 4, 5, 4, 7, 3, 4, 9, 5, 5, 4, 6, 6, 7, 4, 5, 5, 4, 5, 4, 5,
7, 3, 6, 5, 5, 7, 7, 3, 4, 3, 6, 5, 3, 1, 3, 6, 5, 7, 4, 7, 6, 6, 5, 8, 7, 6, 4, 7, 5,
6, 5, 4, 6, 4, 5, 5, 5, 6, 8, 6, 6, 8, 1, 2, 5, 7, 2, 4, 6, 8, 5, 4, 3, 6, 5, 6, 2, 6,
5, 8, 7, 4, 5, 6, 9, 4, 4, 5, 6, 3, 5, 6, 5, 1, 5, 5, 9, 8, 5, 5, 6, 4, 7, 4, 6, 5, 4,
3, 4, 8, 7, 9, 4, 3, 8, 8, 6, 7, 3, 5, 5, 5, 4, 6, 3, 3, 4, 5, 7, 8, 8, 4, 6, 7, 5, 4,
5, 5, 3, 5, 6, 6, 4, 6, 1, 6, 5, 4, 6, 6, 7, 4, 6, 5, 5, 8, 6, 5, 3, 9, 4, 6, 3, 5, 5,
6, 5, 9, 7, 6, 4, 7, 8, 4, 5, 7, 5, 5, 3, 8, 6, 5, 6, 7, 5, 8, 4, 4, 5)

Mind = c(1, 4, 6, 3, 3, 5, 4, 4, 6, 3, 1, 4, 6, 6, 5, 5, 2, 6, 3, 1, 6, 2, 5, 4, 6, 6, 6, 4,
3, 6, 2, 6, 6, 4, 4, 5, 9, 6, 6, 2, 4, 6, 7, 5, 3, 3, 6, 7, 7, 8, 6, 4, 5, 2, 3, 5, 5,
2, 4, 5, 3, 4, 6, 5, 5, 5, 4, 6, 4, 7, 5, 4, 5, 3, 5, 7, 3, 5, 6, 3, 3, 4, 8, 5, 1, 5,
2, 2, 3, 5, 6, 6, 3, 7, 3, 6, 5, 4, 4, 3, 4, 8, 5, 2, 5, 6, 5, 6, 4, 5, 3, 3, 2, 5, 4,
4, 4, 6, 3, 5, 2, 6, 7, 6, 4, 4, 5, 5, 4, 5, 5, 6, 2, 4, 7, 7, 5, 5, 6, 5, 6, 6, 3, 4,
6, 4, 4, 6, 3, 5, 3, 3, 4, 6, 4, 2, 3, 2, 4, 3, 3, 6, 2, 7, 5, 6, 4, 2, 3, 5, 4, 3, 6,
4, 5, 9, 4, 3, 4, 5, 5, 7, 3, 4, 6, 5, 5, 4, 4, 5, 2, 5, 2, 5, 3, 5, 4, 5, 2, 4, 5, 3, 4,
2, 6, 6, 4, 5, 6, 4, 2, 6, 6, 6, 2, 5, 5, 5, 4, 3, 6, 4, 2, 7, 3, 5, 7, 7, 2, 3, 1, 7,
2, 5, 6, 3, 6, 3, 4, 5, 5, 5, 4, 3, 4, 4, 3, 2, 3, 5, 3, 4, 6, 5, 4, 9, 4, 5, 3, 5, 5,
1, 4, 5, 6, 3, 2, 5, 5, 7, 4, 6, 6, 5, 6, 4, 6, 4, 6, 4, 6, 8, 5, 4, 3, 4, 3, 6, 5, 4,
4, 4, 4, 3, 4, 5, 4, 7, 3, 2, 5, 7, 3, 3, 3, 5, 2, 4, 9, 4, 3, 4, 3, 3, 2, 4, 6, 4, 5,
1, 1, 5, 4, 6, 7, 4, 6, 6, 4, 3, 4, 4, 3, 2, 5, 5, 3, 4, 7, 3, 6, 5, 8, 5, 5, 5, 6, 3,
3, 8, 2, 2, 5, 3, 3, 7, 2, 2, 3, 3, 4, 4, 5, 6, 1, 4, 3, 5, 4, 5, 2, 5, 6, 5, 5, 5, 7, 3,
3, 5, 3, 5, 5, 6, 3, 5, 3, 4, 7, 6, 4, 7, 2, 4, 3, 2, 5, 5, 2, 7, 7, 9)

UPM = c(4, 4, 7, 3, 4, 5, 6, 5, 6, 3, 2, 3, 6, 7, 4, 6, 2, 7, 4, 2, 7, 2, 4, 4, 7, 6, 6, 4,
3, 5, 3, 5, 5, 3, 5, 5, 9, 5, 6, 3, 3, 5, 6, 5, 4, 4, 5, 7, 7, 7, 7, 5, 4, 3, 5, 5, 5,

4, 4, 6, 4, 4, 7, 5, 7, 5, 5, 6, 4, 6, 5, 5, 5, 3, 5, 6, 4, 4, 6, 5, 4, 5, 7, 6, 3, 3,
3, 2, 3, 5, 6, 6, 3, 6, 3, 6, 5, 5, 6, 5, 4, 7, 5, 3, 5, 6, 4, 5, 4, 4, 3, 5, 3, 6, 5,
3, 5, 6, 3, 4, 2, 7, 7, 5, 4, 4, 6, 4, 5, 6, 4, 6, 3, 3, 6, 6, 5, 6, 6, 4, 5, 6, 3, 4,
6, 5, 4, 5, 3, 5, 3, 4, 4, 5, 3, 3, 3, 4, 7, 5, 3, 4, 1, 7, 5, 5, 4, 2, 4, 4, 3, 2, 5,
4, 5, 8, 5, 4, 4, 6, 5, 5, 3, 6, 5, 4, 5, 5, 5, 3, 4, 4, 7, 5, 6, 5, 5, 3, 5, 6, 3, 5,
3, 6, 5, 5, 5, 6, 4, 3, 5, 5, 5, 2, 5, 3, 5, 5, 4, 6, 4, 2, 7, 3, 5, 7, 7, 3, 3, 2, 5,
3, 4, 8, 4, 5, 2, 5, 5, 4, 6, 4, 3, 5, 4, 2, 2, 4, 5, 3, 4, 7, 4, 4, 6, 4, 7, 4, 7, 5,
1, 4, 6, 7, 3, 3, 5, 3, 7, 5, 7, 6, 4, 5, 4, 4, 5, 7, 5, 5, 8, 4, 5, 3, 5, 3, 7, 4, 5,
5, 7, 5, 3, 5, 4, 5, 8, 3, 4, 4, 7, 4, 3, 5, 6, 3, 4, 7, 4, 3, 5, 4, 4, 3, 5, 6, 5, 3,
1, 2, 5, 4, 6, 5, 3, 4, 6, 5, 2, 5, 4, 2, 2, 6, 5, 5, 6, 6, 3, 5, 7, 9, 5, 4, 6, 5, 4,
3, 7, 4, 4, 5, 3, 6, 7, 3, 2, 4, 4, 4, 4, 5, 5, 2, 5, 3, 4, 4, 6, 3, 4, 5, 4, 4, 8, 5,
4, 4, 4, 4, 4, 7, 4, 4, 2, 4, 6, 6, 5, 8, 2, 5, 5, 1, 5, 4, 2, 6, 6, 9)

7.5. QUANTILE REGRESSION

Linear regression techniques are designed to help us predict expected values, as in $E(Y) = \mu + \beta X$. But what if our real interest is in predicting extreme values, if, for example, we would like to characterize the observations of Y that are likely to lie in the upper and lower tails of Y's distribution.

Even when expected values or medians lie along a straight line, other quantiles may follow a curved path. Koenker and Hallock applied the method of quantile regression to data taken from Ernst Engel's study in 1857 of the dependence of households' food expenditure on household income. As Figure 7.2 reveals, not only was an increase in food expenditures observed as expected when household income was increased, but the dispersion of the expenditures increased also.

In estimating the τth quantile,[2] we try to find that value of β for which $\Sigma_k \rho_\tau(y_k - f[x_k, \beta])$ is a minimum, where

$$\rho_\tau[x] = \tau x \qquad \text{if } x > 0$$
$$= (\tau - 1)x \quad \text{if } x \leq 0$$

For the 50th percentile or median, we have LAD regression. In fact, we make use of the `rq()` function to obtain the LAD regression coefficients.

Begin by loading the Engel data, which is included in the quantreg library, and plotting the data.

```
➢ library(quantreg)
➢ data(engel)
```

---

[2] τ is pronounced tau.

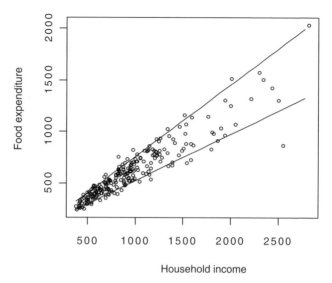

Figure 7.2. Engel data with quantile regression lines superimposed.

```
➢ attach(engel)
➢ plot(x,y,xlab="household income",ylab="food
  expenditure",cex=.5)
```

One of the data points lies at quite a distance from the others. Let us remove it for fear it will have too strong an influence on the resulting regression lines.

```
➢ x1=engel$x[-138]
➢ y1=engel$y[-138]
```

Finally, we overlay the plot with the regression lines for the 10th and 90th percentiles:

```
➢ plot(x1,y1,xlab="household income",ylab="food
  expenditure",cex=.5)
➢ taus = c(.1,.9)
➢  xx = seq(min(x1),max(x1),100)
➢ for(tau in taus){
+   f = coef(rq((y1)~(x1), tau=tau))
+   yy = (f[1]+f[2]*(xx))
+   lines(xx,yy)
+   }
```

Exercise 7.22. Plot the 10th and 90th quantile regression lines for the Engel economic data. How did removing that one data point affect the coefficients?

Exercise 7.23. Brian Cady has collected data relating acorn yield (WTPERHA) to the condition of the oak canopy (OAKCCSI). Plot the 25th, 50th, and 75th quantile regression lines for the following data:

WTPERHA = c(43.04055, 4.946344, 60.21557, 81.66009, 50.61638, 56.65638,
60.92875, 72.11634, 72.80992, 32.66437, 202.776, 116.7866,
14.56848, 59.79613, 37.25652, 93.01928, 20.869, 46.28904,
50.56929, 19.39652, 71.11756, 16.27327, 11.51406, 67.86525,
80.7973, 25.6447, 108.2565, 49.49308, 76.17169, 75.50679,
32.31519, 36.78138, 37.04142, 49.99904, 12.82474, 73.1297,
20.20769, 1.11135, 8.10927, 50.85466, 89.92351, 16.45993,
29.18572)

OAKCCSI = c(0.826136, 0.115109, 0.447214, 0.9, 0.52915, 1, 0.708872,
0.864581, 0.812404, 0.777817, 0.781025, 1, 0.380789, 0.54658,
0.482183, 1, 0.455522, 0.83666, 0.648074, 0.860233, 0.61441,
0.316228, 0.626498, 0.89861, 0.597913, 0.33541, 0.953939,
0.972111, 1, 1, 0.927362, 0.924662, 0.642262, 0.667458, 0.47697,
0.91515, 0.589491, 0.766485, 0.259808, 0.606218, 0.67082,
0.736546, 0.8)

7.6. VALIDATION

As noted in the preceding sections, more than one model can provide a satisfactory fit to a given set of observations; even then, goodness-of-fit is no guarantee of predictive success. Before putting the models we develop to practical use, we need to *validate* them. There are three main approaches to validation:

1. Independent verification (obtained by waiting until the future arrives or through the use of surrogate variables).
2. Splitting the sample (using one part for calibration, the other for verification).
3. Resampling (taking repeated samples from the original sample and refitting the model each time).

In what follows, we examine each of these methods in turn.

7.6.1. Independent Verification

Independent verification is appropriate and preferable whatever the objectives of your model. In geologic and economic studies, researchers often return to the original setting and take samples from points that have been bypassed on the original round. In other studies, verification of the model's form and the choice of variables are obtained by attempting to fit the same model in a similar but distinct context.

For example, having successfully predicted an epidemic at one army base, one would then wish to see if a similar model might be applied at a second and third almost-but-not-quite identical base.

Independent verification can help discriminate among several models that appear to provide equally good fits to the data. Independent verification can be used in conjunction with either of the two other validation methods. For example, an automobile manufacturer was trying to forecast parts sales. After correcting for seasonal effects and long-term growth within each region, ARIMA techniques were used.[3] A series of best-fitting ARIMA models was derived—one model for each of the nine sales regions into which the sales territory had been divided. The nine models were quite different in nature. As the regional seasonal effects and long-term growth trends had been removed, a single ARIMA model applicable to all regions, albeit with coefficients that depended on the region, was more plausible. The model selected for this purpose was the one that gave the best fit when applied to all regions.

Independent verification also can be obtained through the use of surrogate or proxy variables. For example, we may want to investigate past climates and test a model of the evolution of a regional or worldwide climate over time. We cannot go back directly to a period before direct measurements on temperature and rainfall were made, but we can observe the width of growth rings in long-lived trees or measure the amount of carbon dioxide in ice cores.

7.6.2. Splitting the Sample

For validating time series, an obvious extension of the methods described in the preceding section is to hold back the most recent data points, fit the model to the balance of the data, and then attempt to "predict" the values held in reserve.

When time is not a factor, we still would want to split the sample into two parts, one for estimating the model parameters, the other for

---

[3] For examples and discussion of autoregressive integrated moving average (ARIMA) processes used to analyze data whose values change with time, see Brockwell and Davis (1987).

verification. The split should be made at random. The down side is that when we use only a portion of the sample, the resulting estimates are less precise.

In the following exercises, we want you to adopt a compromise proposed by Moiser (1951). Begin by splitting the original sample in half; choose your regression variables and coefficients independently for each of the sub-samples. If the results are more or less in agreement, then combine the two samples and recalculate the coefficients with greater precision.

There are several different ways to program the division in R. Here is one way. Suppose we have 100 triples of observations Y, P1, and P2. Using M and V to denote the values we select for use in estimating and validating, respectively:

```
➢ select=sample(100,50,replace=FALSE)
➢ YM= Y[select]
➢ P1M=P1[select]
➢ P2M=P2[select]
➢ YV=Y[-select]
➢ P1V=P1[-select]
➢ P2V=P2[-select]
```

Exercise 7.24. Apply Moiser's method to the Milazzo data of the previous chapter (Exercise 6.21). Can total coliform levels be predicted on the basis of month, oxygen level, and temperature?

Exercise 7.25. Apply Moiser's method to the data provided in Exercises 7.20 and 7.21 to obtain prediction equation(s) for Attitude in terms of some subset of the remaining variables.

Note: As conditions and relationships do change over time, any method of prediction should be *revalidated* frequently. For example, suppose we had used observations from January 2000 to January 2004 to construct our original model, and held back more recent data from January to June 2004 to validate it. When we reach January 2005, we might refit the model, using the data from 1/2000 to 6/2004 to select the variables and determine the values of the coefficients, then use the data from 6/2004 to 1/2005 to validate the revised model.

Exercise 7.26. Some authorities would suggest discarding the earliest observations before refitting the model. In the present example, this would mean discarding all the data from the first half of the year 2000. Discuss the possible advantages and disadvantages of discarding these data.

7.6.3. Cross-validation with the Bootstrap

Recall that the purpose of bootstrapping is to simulate the taking of repeated samples from the original population (and to save money and time by not having to repeat the entire sampling procedure from scratch). By bootstrapping, we are able to judge to a limited extent whether the models we derive will be useful for predictive purposes or whether they will fail to carry over from sample to sample. As the next exercise demonstrates, some variables may prove more reliable as predictors than others.

Exercise 7.27. Bootstrap repeatedly from the data provided in Exercises 7.20 and 7.21 and use the R **step()** function to select the variables to be incorporated in the model each time. Are some variables common to all the models?

7.7. CLASSIFICATION AND REGRESSION TREES (CART)

As the number of potential predictors increases, the method of linear regression becomes less and less practical. With three potential predictors, we can have as many as seven coefficients to be estimated: one for the intercept, three for first-order terms in the predictors P_i, two for second-order terms of the form P_iP_j, and one third-order term $P_1P_2P_3$. With k variables, we have k first-order terms, $k(k-1)$ second-order terms, and so forth. Should all these terms be included in our model? Which ones should be neglected? With so many possible combinations, will a single equation be sufficient?

We need to consider alternate approaches. If you're a mycologist, a botanist, a herpetologist, or simply a nature lover you may have made use of some sort of a key. For example,

I. Leaves simple?
   3. Leaves needle-shaped?
      a. Leaves in clusters of 2 to many?
         i. Leaves in clusters of 2 to 5, sheathed, persistent for several years?

The R **tree()** procedure attempts to construct a similar classification scheme from the data you provide. You'll need to install a new package of R functions in order to do the necessary calculations. The easiest way to do this is get connected to the internet and then type

```
➤ install.packages("tree")
```

The installation, which includes downloading, unzipping, and integrating the new routines, is then done automatically. Installation needs to be done once and once only. But each time, you'll need to load the "tree" functions into computer memory before you can use the CART routines by typing

```
➤ library(tree)
```

Figure 7.3 is an example of a tree we constructed using the data from the consumer survey and the R commands

```
➤ catt.tr=tree(Purchase ~ Fashion + Ozone + Gamble)
➤ plot(catt.tr,type="u"); text(catt.tr,srt=90)
```

Note that some of the decisions seemed to be made on the basis of a single predictor—Gamble; others utilized the values of two predictors—Gamble and Fashion; and still others utilized the values of Gamble, Fashion, and Ozone, the equivalent of three separate linear regression models.

Exercise 7.28. Show that the R function **tree()** only makes use of variables it considers important by constructing a regression tree based on the model

```
Purchase~Fashion+Ozone+Pollution+Coupons+Gamble+Today+Work+UPM
```

Exercise 7.29. Apply the CART method to the Milazzo data of the previous chapter (Exercise 6.21) to develop a prediction scheme for

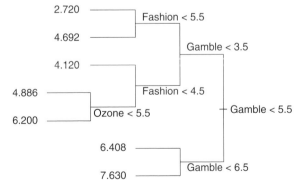

Figure 7.3. Labeled regression tree for predicting purchase attitude.

coliform levels in bathing water based on the month, oxygen level, and temperature.

The output from our example and from Exercises 7.26 and 7.27 are *regression trees*, that is, at the end of each *terminal node* is a numerical value. While this might be satisfactory in an analysis of our Milazzo coliform data, it is misleading in the case of our consumer survey where our objective is to pinpoint the most likely sales prospects.

Suppose instead that we begin by grouping our customer attitudes into categories. Purchase attitudes of 1, 2, or 3 indicate low interest, 4, 5, and 6 indicate medium interest, and 7, 8, and 9 indicate high interest. To convert them using R, we enter

```
➢ Pur.cat = cut(Purchase,breaks=c(0,3,6,9),labels=c("lo",
  "med","hi"))
```

Now, we can construct the classification tree depicted in Figure 7.4.

```
➢ catt.tr=tree(Pur.cat ~ Fashion + Ozone + Gamble)
➢ plot(catt.tr,type="u"); text(catt.tr,srt=90)
```

Many of the branches of this tree appear redundant. If we have already classified a prospect, there is no point in additional questions. We can remove the excess branches and create a display similar to Figure 7.5 with the following code:

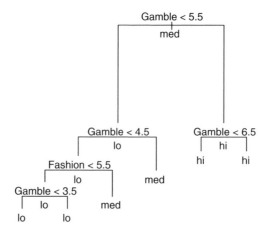

Figure 7.4. Labeled classification tree for predicting prospect attitude.

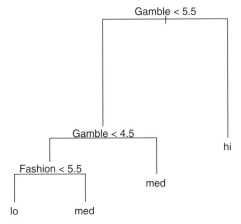

Figure 7.5. Pruned classification tree for predicting prospect attitude.

```
➢ tmp=prune.tree(catt.tr, best=4)
➢ plot(tmp); text(tmp)
```

These classifications are actually quite tentative; typing the command

```
➢ print(tmp)
```

will display the nodes in this tree along with the proportions of correct and incorrect assignments.

```
node), split, n, deviance, yval, (yprob)
  * denotes terminal node
1) root 400 878.70 med (0.3225 0.3375 0.3400)
  2) Gamble < 5.5 275 567.60 lo (0.1782 0.4618 0.3600)
  4) Gamble < 4.5 192 372.30 lo (0.1615 0.5677 0.2708)
      8) Fashion < 5.5 160 281.00 lo (0.1187 0.6438
         0.2375) *
      9) Fashion > 5.5 32  66.77 med (0.3750 0.1875
         0.4375) *
  5) Gamble > 4.5 83 163.50 med (0.2169 0.2169
     0.5663) *
  3) Gamble > 5.5 125 205.50 hi (0.6400 0.0640
     0.2960) *
```

The three fractional values within parentheses refer to the proportions in the data set that were assigned to the "hi," "lo," and "med" categories, respectively. For example, in the case when Gamble > 5.5, 64% are assigned correctly by the classification scheme, while just over 6% who are actually poor prospects (with scores of 3 or less) are assigned in error to the high category.

Exercise 7.30. Divide the consumer survey data into an estimation and a validation sample.

(a) Create a classification tree for Purchase as a function of *all* the variables in the consumer survey. Treat each Purchase level as a single category—this can be done in a single step by typing `Pur.fact= factor(Purchase)`. Prune the tree if necessary.

(b) Compare your findings with those you obtained in Exercise 7.21 via multiple regression techniques.

(c) Validate the tree you built using the `predict()` function located in the tree library.

7.8. SUMMARY AND REVIEW

In this chapter, you were introduced to two techniques for classifying and predicting outcomes: linear regression and classification and regression trees. Three methods for estimating linear regression coefficients were described along with guidelines for choosing among them. You were provided with a stepwise technique for choosing variables to be included in a regression model. The assumptions underlying the regression technique were discussed along with the resultant limitations. Overall guidelines for model development were provided.

You learned the importance of validating and revalidating your models before placing any reliance upon them. You were introduced to one of the simplest of pattern recognition methods, the classification and regression tree to be used whenever there are large numbers of potential predictors or when classification rather than quantitative prediction is your primary goal.

You learned how to download and install library packages not part of the original R download including `library(quantreg)` and `library (tree)`. You made use of several modeling functions including `lsfit`, `lm` (with its associated functions `fitted`, `summary`, `resid`, and `step`), `rq` (with its associated function `coeff`), and `tree` (with its associated function `prune.tree`). You learned two new ways to fill an R vector and to delete selected elements.

Exercise 7.31. Make a list of all the italicized terms in this chapter. Provide a definition for each one along with an example.

Exercise 7.32. It is almost self-evident that levels of toluene, a commonly used solvent, would be observed in the blood after working in a room where the solvent was present in the air. Do the observations recorded below also suggest that blood levels are a function of age and body weight? Construct a model for predicting blood levels of toluene using this data.

Blood toluene	0.494	0.763	0.534	0.552	1.084	0.944	0.955	0.696
Air toluene	50	50	50	50	100	100	100	100
Weight	378	439	302	405	421	370	363	389
Age	95	95	84	85	86	86	83	86

Blood toluene	12.085	9.647	7.524	10.783	38.619	25.402	26.481	28.155
Air toluene	500	500	500	500	1000	1000	1000	1000
Weight	371	347	misg	348	378	433	363	420
Age	83	84	85	85	93	93	85	86

Exercise 7.33. Using the data from Exercise 6.19, develop a model for predicting whether an insurance agency will remain solvent.

Exercise 7.34. The weights of rat fetuses sacrificed at various intervals after conception are recorded below. Test the hypothesis that the weight of a rat fetus doubles every 2.5 time intervals.

Interval	1	2	2	3	4	5	5	6	7	9	9
Weight	2.44	4.46	4.00	2.21	10.8	10.4	10.13	15.78	15.50	13.2	16.6

8

REPORTING YOUR FINDINGS

In this chapter, we assume you have just completed an analysis of your own or someone else's research and now wish to issue a report on the overall findings. You'll learn what to report and how to go about reporting it with particular emphasis on the statistical aspects of data collection and analysis.

One of the most common misrepresentations in scientific work is the scientific paper itself. It presents a mythical reconstruction of what actually happened. All of what are in retrospect mistaken ideas, badly designed experiments and incorrect calculations are omitted. The paper presents the research as if it had been carefully thought out, planned and executed according to a neat, rigorous process, for example, involving testing of a hypothesis. The misrepresentation of the scientific paper is the most formal aspect of the misrepresentation of science as an orderly process based on a clearly defined method.
—Brian Martin

Introduction to Statistics Through Resampling Methods and R/S-PLUS®, By Phillip I. Good
Copyright © 2005 by John Wiley & Sons, Inc.

8.1. WHAT TO REPORT

Reportable elements include all of the following:

- Study objectives
- Hypotheses
- Power and sample size calculations
- Data collection methods
- Validation methods
- Data summaries
- Details of the statistical analysis
- Sources of missing data
- Exceptions

Study Objectives. If you are contributing to the design or analysis of someone else's research efforts, a restatement of the objectives is an essential first step. This ensures that you and the principal investigator are on the same page. This may be necessary in order to formulate quantifiable, testable hypotheses.

Objectives may have shifted or been expanded upon. Often such changes are not documented. You cannot choose or justify the choice of statistical procedures without a thorough understanding of study objectives.

Hypotheses. To summarize what was stated in Chapter 5, both your primary and alternative hypotheses must be put in quantified testable form. Your primary hypothesis is used to establish the significance level and your alternative hypothesis to calculate the power of your tests.

Your objective may be to determine whether adding a certain feature to a product would increase sales. (Quick: Will this alternative hypothesis lead to a one-sided or a two-sided test?) Yet for reasons that have to do solely with the limitations of statistical procedures, your primary hypothesis will normally be a null hypothesis of no effect.

Not incidentally, as we saw in Chapter 6, the optimal statistical test for an ordered response is quite different from the statistic one uses for detecting an arbitrary difference among approaches. All the more reason why we need to state our alternative hypotheses explicitly.

Power and Sample Size Calculations. Your readers will want to know the details of your power and sample size calculations early on. If you don't let them know, they may assume the worst, for example, that your sample size is too small and your survey is not capable of detecting significant effects.

State the alternative hypotheses that your study is intended to detect. Reference your methodology and/or the software you employed in making your power calculations. State the source(s) you relied on for your initial estimates of incidence and variability.

Here is one example. "Over a ten year period in the Himalayas, Dempsy and Peters (1995) observed an incidence of five infected individuals per 100 persons per year. To ensure a probability of at least 90% of detecting a reduction in disease incidence from five persons to one person per 100 persons per year using a one-sided Fisher's exact test at the 2.5% significance level, 400 individuals were assigned to each experimental procedure group. This sample size was determined using the StatXact-5 power calculations for comparing two binomials."

Data Collection Methods. Although others may have established the methods of data collection, a comprehensive knowledge of these methods is essential to your choice of statistical procedures and should be made apparent in report preparation. Consider that 90% of all errors occur during data collection as observations are erroneously recorded (GIGO), guessed at, or even faked. Seemingly innocuous work-arounds may have jeopardized the integrity of the study. You need to know and report on exactly how the data were collected, not on how they were supposed to have been collected.

You need to record how study subjects were selected, what was done to them (if appropriate), and when and how this was done. Details of recruitment or selection are essential if you are to convince readers that your work is applicable to a specific population. If incentives were used (phone cards, T-shirts, cash), their use should be noted.

Readers will want to know the nature and extent of any blinding (and of the problems you may have had to overcome to achieve it). They will want to know how each observational subject was selected—random, stratified, or cluster sampling? They will want to know the nature of the controls (and the reasoning underlying your choice of a passive or active control experimental procedure) and of the experimental design. Did each subject act as his own control as in a crossover design? Were case controls employed? If they were matched, how were they matched?

You will need the answers to all of the following questions and should incorporate each of the answers in your report:

- What was the survey or experimental unit?
- What were all the potential sources of variation?
- How was each of the individual sources compensated for? In particular, was the sample simple or stratified?

- How were subjects or units grouped into strata?
- Were the units sampled in clusters or individually?
- How were subjects assigned to experimental procedures? If assignment was at random, how was this accomplished?
- How was independence of the observations assured?

Clusters. Surveys often take advantage of the cost savings that result from naturally occurring groups such as work sites, schools, clinics, neighborhoods, even entire towns or states. Not surprisingly, the observations within such a group are correlated. For example, individuals in the same community or work group often have shared views. Married couples, the ones whose marriages last, tend to have shared values. The effect of such correlation must be accounted for by the use of the appropriate statistical procedures. Thus, the nature and extent of such cluster sampling must be spelled out in detail in your reports.

Exercise 8.1. Examine three recent reports in your field of study or in any field that interests you. (Examine three of your own reports if you have them.) Answer the following in each instance:

(a) What was the primary hypothesis?

(b) What was the principal alternative hypothesis?

(c) Which statistics were employed and why?

(d) Were all the assumptions underlying this statistic satisfied?

(e) Was the power of this statistic reported?

(f) Was the statistic the most powerful available?

(g) If significant, was the size of the effect estimated?

(h) Was a one-tailed or two-tailed test used? Was this correct?

(i) What was the total number of tests that were performed?

If the answers to these questions were not present in the reports you reviewed, what was the reason? Had their authors something to hide?

Validation Methods. A survey will be compromised if any of the following is true:

- Participants are not representative of the population of interest.
- Responses are not independent among respondents.
- Nonresponders, that is, those who decline to participate, would have responded differently.

· Respondents lie or answer carelessly.
· Forms are incomplete.

Your report should detail the preventive measures employed by the investigator and the degree to which they were successful.

You should describe the population(s) of interest in detail—providing demographic information where available—and similarly characterize the samples of participants to see whether they are indeed representative. (Graphs are essential here.)

You should describe the measures taken to ensure responses were independent, including how participants were selected and where and how the survey took place.

A sample of nonrespondents should be contacted and evaluated. The demographics of the nonrespondents and their responses should be compared with those of the original sample.

How do you know respondents told the truth? You should report the results of any crosschecks such as redundant questions. And you should report the frequency of response omissions on a question-by-question basis.

8.2. TEXT, TABLE, OR GRAPH?

Whatever is the use of a book without pictures?
—Alice in *Alice in Wonderland*

A series of major decisions need to be made as to how you will report your results—text, table, or graph? Whatever Alice's views on the subject, a graph may or may not be more efficient at communicating numeric information than the equivalent prose. This efficiency is in terms of the amount of information successfully communicated and not necessarily any space savings. Resist the temptation to enhance your prose with pictures.

And don't fail to provide a comprehensive caption for each figure. As Good and Hardin note in their text, *Common Errors in Statistics* (Wiley, 2003), if the graphic is a summary of numeric information, then the graph caption is a summary of the graphic. The text should be considered part of the graphic design and should be carefully constructed rather than placed as an afterthought. Readers, for their own use, often copy graphics and tables that appear in articles and reports. A failure on your part to completely document the graphic in the caption can result in gross misrepresentation in these cases.

A sentence should be used for displaying two to five numbers, as in "The blood type of the population of the United States is approximately 45% O,

40% A, 11% B, and 4% AB." Note that the blood types are ordered by frequency.

Tables with appropriate marginal means are often the best method of presenting results. Consider adding a row (or column, or both) of contrasts; for example, if the table has only two rows we could add a row of differences, row 1 minus row 2.

Tables dealing with two-factor arrays are straightforward, provided confidence limits are clearly associated with the correct set of figures. Tables involving three or more factors simultaneously are not always clear to the reader and are best avoided.

Make sure the results are expressed in appropriate units. For example, parts per thousand may be more natural than percentages in certain cases.

A table of deviations from row and column means (or tables, if there are several strata) can alert us to the presence of outliers and may also reveal patterns in the data that were not yet considered.

Exercise 8.2. To report each of the following, should you use text, a table, or a graph? If a graphic, then what kind?

- Number of goals (each of five teams)
- Blood types of Australians
- Comparison treated/control red blood cell counts
- Comparison of blood types in two populations
- Location of leukemia cases by county
- Arm span versus height (six persons)

8.3. SUMMARIZING YOUR RESULTS

Your objective in summarizing your results should be to communicate some idea of all of the following:

- The populations from which your samples are drawn.
- Your estimates of population parameters.
- The dispersion of the distribution from which the observations were drawn.
- The precision of your estimates.

Proceed in three steps: First, characterize the populations and subpopulations from which your observations are drawn. Of course, this is the main goal in studies of market segmentation. A histogram or scatter plot can help communicate the existence of such subpopulations to our readers. Few real-life distributions resemble the bell-shaped normal curve depicted in Figure 1.7. Most are bi- or even trimodal with each mode or peak corresponding to a distinct subpopulation. We can let the histogram speak for itself; but a better idea, particularly if you already suspect that the basis for market segments is the value of a second variable (such as home ownership or level of education), is to add an additional dimension by dividing each of the histogram's bars into differently shaded segments whose size corresponds to the relative numbers in each subpopulation. Our R program to construct two overlaid histograms makes use of data frames, selection by variable value [hts$sex= ="b",], and the options of the **hist()** function. Figure 8.1 is the result.

```
➢ height = c(141, 156.5, 162, 159, 157, 143.5, 154, 158,
   140, 142, 150, 148.5, 138.5, 161, 153, 145, 147, 158.5,
   160.5, 167.5, 155, 137)
➢ sex = c("b",rep("g",7), "b",rep("g",6),rep("b",7))
➢ hts = data.frame(sex,height)
➢ hist(hts[hts$sex=="b",]$height,
+ xlim=c(min(height), max(height)), ylim=c(0,4),
+ col="blue")
➢ hist(hts[hts$sex=="g",]$height,
```

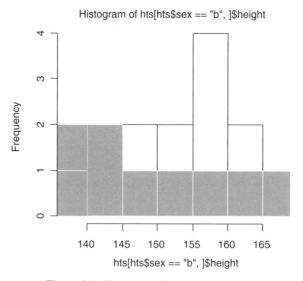

Figure 8.1. Histograms of class data by sex.

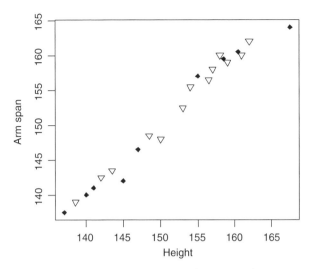

Figure 8.2. Overlying scatter plots of class data by sex.

```
+ xlim=c(min(height), max(height)), ylim=c(0,4),
+ add=T, border="pink")
```

Similarly, we can provide for different subpopulations on a two-dimensional scatter plot by using different colors or shapes for the points.

```
➤ armspan = c(141, 156.5, 162, 159, 158, 143.5, 155.5,
  160, 140, 142.5, 148, 148.5, 139, 160, 152.5, 142,
  146.5, 159.5, 160.5, 164, 157, 137.5)
➤ psex = c(18,rep(25,7),18,rep(25,6),rep(18,7))
➤ plot(height,armspan, pch=psex)
```

yields Figure 8.2, while

```
➤ coplot(height~armspan|sexf)
➤ plot(sexf,height)
```

produces Figure 8.3.

8.3.1. Center of the Distribution

For small samples of three to five observations, summary statistics are virtually meaningless. Reproduce the actual observations; this is easier to do and more informative.

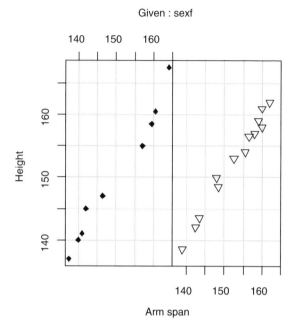

Figure 8.3. Side-by-side scatter plots of class data by sex.

Though the arithmetic mean or average is in common use, it can be very misleading. For example, the mean income in most countries is far in excess of the *median* income or 50th percentile to which most of us can relate. When the arithmetic mean is meaningful, it is usually equal to or close to the median. Consider reporting the median in the first place.

The *geometric mean* is more appropriate than the arithmetic in three sets of circumstances:

1. When losses or gains can best be expressed as a percentage rather than a fixed value.
2. When rapid growth is involved.
3. When the data span several orders of magnitude, as with the concentration of pollutants.

Because bacterial populations can double in number in only a few hours, many government health regulations utilize the geometric rather than the arithmetic mean. A number of other government regulations also use it though the sample median would be far more appropriate (and understandable). In any event, since your average reader may be unfamiliar with the geometric mean be sure to comment on its use and on your reasons for adopting it.

The purpose of your inquiry must be kept in mind. Orders (in $) from a machinery plant ranked by size may be quite skewed with a few large orders. The median order size might be of interest in describing sales; the mean order size would be of interest in estimating revenues and profits.

Are the results expressed in appropriate units? For example, are parts per thousand more natural in a specific case than percentages? Have we rounded off to the correct degree of precision, taking account of what we know about the variability of the results, and considering whether the reader will use them, perhaps by multiplying by a constant factor, or another variable?

Whether you report a mean or a median, be sure to report only a sensible number of decimal places. Most statistical packages, including R, can give you nine or ten. Don't use them. If your observations were to the nearest integer, your report on the mean should include only a single decimal place. Limit tabulated values to no more than two effective (changing) digits. Readers can distinguish 354,691 and 354,634 at a glance but will be confused by 354,691 and 357,634.

8.3.2. Dispersion

The standard error of a summary is a useful measure of uncertainty *if* the observations come from a normal or Gaussian distribution (see Figure 1.7). Then in 95% of the samples we would expect the sample mean to lie within two standard errors of the population mean.

But if the observations come from any of the following:

- a nonsymmetric distribution like an exponential or a Poisson
- a truncated distribution like the uniform
- a mixture of populations

we *cannot* draw any such inference. For such a distribution, the probability that a future observation would lie between plus and minus one standard error of the mean might be anywhere from 40% to 100%.

Recall that the standard error of the mean equals the standard deviation of a single observation divided by the square root of the sample size. As the standard error depends on the squares of individual observations, it is particularly sensitive to outliers. A few extra large observations, even a simple typographical error, might have a dramatic impact on its value.

If you can't be sure your observations come from a normal distribution, then for samples from nonsymmetric distributions of size 6 or less, tabulate the minimum, the median, and the maximum. For samples of size 7 and up, consider using a box and whiskers plot. For samples of size 30 and up, the bootstrap may provide the answer you need.

8.4. REPORTING ANALYSIS RESULTS

How you conduct and report your analysis will depend on whether or not

- Baseline results of the various groups are equivalent.
- Results of the disparate experimental procedure sites may be combined (if multiple observations sites were used).
- Results of the various adjunct experimental procedure groups may be combined (if adjunct or secondary experimental procedures were used).
- Missing data, dropouts, and withdrawals are unrelated to experimental procedure.

Thus, your report will have to include:

1. Demonstrations of similarities and differences for the following:
   - Baseline values of the various experimental procedure groups.
   - End points of the various subgroups determined by baseline variables and adjunct therapies.
2. Explanations of protocol deviations including:
   - Ineligibles who were accidentally included in the study.
   - Missing data.
   - Dropouts and withdrawals.
   - Modifications to procedures.

Further explanations and stratifications will be necessary if the rates of any of the above protocol deviations differ among the groups assigned to the various experimental procedures. For example, if there are differences in the baseline demographics, then subsequent results will need to be stratified accordingly. Moreover, some plausible explanation for the differences must be advanced.

Here is an example. Suppose the vast majority of women in the study were in the control group. To avoid drawing false conclusions about the men, the results for men and women must be presented separately, unless one first can demonstrate that the experimental procedures have similar effects on men and women.

Report the results for each primary end point separately. For each end point:

(a) Report the aggregate results by experimental procedure for all who were examined during the study for whom you have end-point or intermediate data.

(b) Report the aggregate results by experimental procedure only for those subjects who were actually eligible, who were treated originally as randomized, or who were not excluded for any other reason. Provide significance levels for comparisons of experimental procedures.

(c) Break down these latter results into subsets based on factors determined prior to the start of the study as having potential impact on the response to treatment, such as adjunct therapy or gender. Provide significance levels for comparisons of experimental procedures for these subsets of cases.

(d) List all factors uncovered during the trials that appear to have altered the effects of the experimental procedures. Provide a tabular comparison by experimental procedure for these factors but do *not* include *p*-values. The probability calculations that are used to generate *p*-values are not applicable to hypotheses and subgroups that are conceived *after* the data have been examined.

If there are multiple end points, you have the option of providing a further multivariate comparison of the experimental procedures.

Last, but by no means least, you must report the number of tests performed. When we perform multiple tests in a study, there may not be room (nor interest) in which to report all the results, but we do need to report the total number of statistical tests performed so that readers can draw their own conclusions as to the significance of the results that are reported. To repeat a finding of previous chapters, when we make 20 tests at the 1 in 20 or 5% significance level, we expect to find at least one or perhaps two results that are "statistically significant" by chance alone.

8.4.1. *p*-Values or Confidence Intervals?

As you read the literature of your chosen field, you will soon discover that *p*-values are more likely to be reported than confidence intervals. We don't agree with this practice and here is why.

Before we perform a statistical test, we are concerned with its significance level, that is, the probability that we will mistakenly reject our hypothesis when it is actually true. In contrast to the significance level, the *p*-value is a random variable that varies from sample to sample. There may be highly significant differences between two populations and yet the samples taken from those populations and the resulting *p*-value may not reveal that difference. Consequently, it is not appropriate for us to compare the *p*-values from two distinct experiments, or from tests on two variables measured in the same experiment, and declare that one is more significant than the other.

If we agree in advance of examining the data that we will reject the hypothesis if the *p*-value is less than 5%, then our significance level is 5%. Whether our *p*-value proves to be 4.9% or 1% or 0.001%, we will come to the same conclusion. One set of results is not more significant than another; it is only that the difference we uncovered was measurably more extreme in one set of samples than in another.

We are less likely to mislead and more likely to communicate all the essential information if we provide confidence intervals about the estimated values. A confidence interval provides us with an estimate of the size of an effect as well as telling us whether an effect is significantly different from zero.

Confidence intervals, you will recall from Chapter 4, can be derived from the rejection regions of our hypothesis tests. Confidence intervals include all values of a parameter for which we would accept the hypothesis that the parameter takes that value.

Warning: A common error is to misinterpret the confidence interval as a statement about the unknown parameter. It is not true that the probability that a parameter is included in a 95% confidence interval is 95%. Nor is it at all reasonable to assume that the unknown parameter lies in the middle of the interval rather than toward one of the ends. What is true is that if we derive a large number of 95% confidence intervals, we can expect the true value of the parameter to be included in the computed intervals 95% of the time. Like the *p*-value, the upper and lower confidence limits of a particular confidence interval are random variables, for they depend on the sample that is drawn.

The probability the confidence interval covers the true value of the parameter of interest and the method used to derive the interval must both be reported.

Exercise 8.3. Give at least two examples to illustrate why *p*-values are not applicable to hypotheses and subgroups that are conceived after the data is examined.

8.5. EXCEPTIONS ARE THE REAL STORY

Before you draw conclusions, be sure you have accounted for all missing data, interviewed nonresponders, and determined whether the data were missing at random or were specific to one or more subgroups.

Let's look at two examples, the first involving nonresponders, the second airplanes.

8.5.1. Nonresponders

A major source of frustration for researchers is when the variances of the various samples are unequal. Alarm bells sound. *t*-Tests and the analysis of variance are no longer applicable; we run to the textbooks in search of some variance-leveling transformation. And completely ignore the phenomena we've just uncovered.

If individuals have been assigned at random to the various study groups, the existence of a significant difference in any parameter suggests that there is a difference in the groups. The primary issue is to understand why the variances are so different, and what the implications are for the subjects of the study. It may well be the case that a new experimental procedure is not appropriate because of higher variance, even if the difference in means is favorable. This issue is important whether or not the difference was anticipated.

In many clinical measurements there are minimum and maximum values that are possible. If one of the experimental procedures is very effective, it will tend to push patient values into one of the extremes. This will produce a change in distribution from a relatively symmetric one to a skewed one, with a corresponding change in variance.

The distribution may not be unimodal. A large variance may occur because an experimental procedure is effective for only a subset of the patients. Then you are comparing mixtures of distributions of responders and nonresponders; specialized statistical techniques may be required.

8.5.2. The Missing Holes

During the Second World War, a group was studying planes returning from bombing Germany. They drew a rough diagram showing where the bullet holes were and recommended those areas be reinforced. Abraham Wald, a statistician, pointed out that essential data were missing. What about the planes that didn't return?

When we think along these lines, we see that the areas of the returning planes that had almost no apparent bullet holes have their own story to tell.

Bullet holes in a plane are likely to be at random, occurring over the entire plane. The planes that did not return were those that were hit in the areas where the returning planes had no holes. Do the data missing from your own experiments and surveys also have a story to tell?

8.5.3. Missing Data

As noted in an earlier section of this chapter, you need to report the number and source of all missing data. But especially important is to summarize and describe all those instances in which the incidence of missing data varied among the various treatment and procedure groups.

Here are two examples where the missing data was the real finding of the research effort.

To increase participation, respondents to a recent survey were offered a choice of completing a printed form or responding online. An unexpected finding was that the proportion of missing answers from the online survey was half that from the printed forms.

A minor drop in cholesterol levels was recorded among the small fraction of participants who completed a recent trial of a cholesterol-lowering drug. As it turned out, almost all those who completed the trial were in the control group. The numerous dropouts from the treatment group had only unkind words for the test product's foul taste and undrinkable consistency.

8.5.4. Recognize and Report Biases

Very few studies can avoid bias at some point in sample selection, study conduct, and results interpretation. We focus on the wrong end points; participants and coinvestigators see through our blinding schemes; the effects of neglected and unobserved confounding factors overwhelm and outweigh the effects of our variables of interest. With careful and prolonged planning, we may reduce or eliminate many potential sources of bias, but seldom will we be able to eliminate all of them. Accept bias as inevitable and then endeavor to recognize and report all that do slip through the cracks.

Most biases occur during data collection, often as a result of taking observations from an unrepresentative subset of the population rather than from the population as a whole. An excellent example is the study that failed to include planes that did *not* return from combat.

When analyzing extended seismological and neurological data, investigators typically select specific cuts (a set of consecutive observations in time) for detailed analysis, rather than trying to examine all the data (a near impossibility). Not surprisingly, such "cuts" usually possess one or more intriguing features not to be found in run-of-the-mill samples. Too often theories evolve from these very biased selections.

The same is true of meteorological, geological, astronomical, and epidemiological studies, where with a large amount of available data, investigators naturally focus on the "interesting" patterns.

Limitations in the measuring instrument such as censoring at either end of the scale can result in biased estimates. Current methods of estimating cloud optical depth from satellite measurements produce biased results that depend strongly on satellite viewing geometry. Similar problems arise in high-temperature and high-pressure physics and in radioimmunoassay. In psychological and sociological studies, too often we measure that which is convenient to measure rather than that which is truly relevant.

Close collaboration between the statistician and the domain expert is essential if all sources of bias are to be detected and, if not corrected, accounted for and reported. We read a report recently by economist Otmar Issing in which it was stated that the three principal sources of bias in the measurement of price indices are substitution bias, quality change bias, and new product bias. We've no idea what he was talking about, but we do know that we would never attempt an analysis of pricing data without first consulting an economist.

8.6. SUMMARY AND REVIEW

In this chapter, we discussed the necessary contents of your reports whether on your own work or that of others. We reviewed what to report, the best form in which to report it, and the appropriate statistics to use in summarizing your data and your analysis. We also discussed the need to report sources of missing data and potential biases.

9

PROBLEM SOLVING

If you have made your way through the first eight chapters of this text, then you may already have found that more and more people, strangers as well as friends, are seeking you out for your newly acquired expertise. (Not as many as if you were stunningly attractive or a film star, but a great many people nonetheless.) Your boss may even have announced that from now on you will be the official statistician of your group.

To prepare you for your new role in life, you will be asked in this chapter to work your way through a wide variety of problems that you may well encounter in practice. A final section will provide you with some over all guidelines. You'll soon learn that deciding which statistic to use is only one of many decisions that need be made.

9.1. THE PROBLEMS

1. With your clinical sites all lined up and everyone ready to proceed with a trial of a new experimental vaccine versus a control, the manufacturer tells you that because of problems at the plant, the 10,000 ampoules of vaccine you've received are all he will be able to send you. Explain why you can no longer guarantee the power of the test.

Introduction to Statistics Through Resampling Methods and R/S-PLUS®, By Phillip I. Good
Copyright © 2005 by John Wiley & Sons, Inc.

2. After collecting some 50 observations, 25 on members of a control group, and 25 who have taken a low dose of a new experimental drug, you decide to add a third high-dose group to your clinical trial, and to take 75 additional observations, 25 on the members of each group. How would you go about analyzing these data?

3. You are given a data sample and asked to provide an interval estimate for the population variance. What two questions ought you to ask about the sample first?

4. John would like to do a survey on the use of controlled substances by teenagers but realizes he is unlikely to get truthful answers. He comes up with the following scheme: Each respondent is provided with a coin, instructions, a question sheet containing two questions, and a sheet on which to write his/her answer, yes or no. The two questions are:

(a) Is a cola (Coke or Pepsi) your favorite soft drink? Yes or No?

(b) Have you used marijuana within the past seven days? Yes or No?

The teenaged respondents are instructed to flip the coin so that the interviewer cannot see it. If the coin comes up heads, they are to write their answer to the first question on the answer sheet; otherwise they are to write their answer to question 2.

Show that this approach will be successful, providing John already knows the proportion of teenagers who prefer colas to other types of soft drinks.

5. The town of San Philippe has asked you to provide confidence intervals for the recent census figures for their town. Are you able to do so? Could you do so if you had some additional information? What might this information be? Just how would you go about calculating the confidence intervals?

6. The town of San Philippe has called on you once more. They have in hand the annual income figures for the past six years for their town and for their traditional rivals at Carfad-Sur-La-Mere and want you to make a statistical comparison. Are you able to do so? Could you do so if you had some additional information? What might this information be? Just how would you go about calculating the confidence intervals?

7. You have just completed your analysis of a clinical trial and have found a few minor differences between patients subjected to the standard and revised procedures. The marketing manager has gone over your findings and noted that the differences are much greater if limited to patients who passed their first post procedure day without complications. She asks you for a p-value. What do you reply?

8. At the time of his death in 1971, psychologist Cyril Burt was viewed as an esteemed and influential member of his profession. Within months, psychologist Leon Kamin reported numerous flaws in Burt's research

involving monozygotic twins who were reared apart. Shortly thereafter, a third psychologist, Arthur Jensen, also found fault with Burt's data.

Their primary concern was the suspicious consistency of the correlation coefficients for the intelligence test scores of the monozygotic twins in Burt's studies. In each study Burt reported sum totals for the twins he had studied so far. His original results were published in 1943. In 1955 he added six pairs of twins and reported results for a total of 21 sets of twins. Likewise in 1966 he reported the results for a total of 53 pairs. In each study Burt reported correlation coefficients indicating the similarity of intelligence scores for monozygotic twins who were reared apart. A high correlation coefficient would make a strong case for Burt's hereditarian views.

Burt reported the following coefficients: 1943, $r = .770$; 1955, $r = .771$; 1966, $r = .771$. Why was this suspicious?

9. Which hypothesis testing method—permutation, parametric, or bootstrap—would you use to address each of the following?

(a) Testing for an ordered dose response.

(b) Testing whether the mean time to failure of new light bulb in intermittent operation is one year.

(c) Comparing two drugs using the data from the following contingency table.

	Drug A	Drug B
Respond	5	9
No	5	1

(d) Comparing old and new procedures using the data from the following 2×2 factorial design:

	Control	Old
Control		1,150
		2,520
		900
		50
Young	5,640	7,100
	5,120	11,020
	780	13,065
	4,430	
	7,230	

ETHICAL STANDARD

Polish-born Jerzy Neyman (1894–1981) is generally viewed as one of the most distinguished statisticians of the 20th century. Along with Egon Pearson, he is responsible for the method of assigning the outcomes of a set of observations to either an acceptance or a rejection region in such a way that the power is maximized against a given alternative at a specified significance level. He was asked by the United States government to be part of an international committee monitoring the elections held in a newly liberated Greece after World War II. In the oversimplified view of the U.S. State Department, there were two groups running in the election: the Communists and the Good Guys. Professor Neyman's report that both sides were guilty of extensive fraud pleased no one but set an ethical standard for other statisticians to follow.

10. The government has just audited 200 of your company's submissions over a four-year period and has found that the average claim was in error in the amount of $135. Multiplying $135 by the 4000 total submissions during that period, they are asking your company to reimburse them in the amount of $540,000. List all possible objections to the government's approach.

11. Since I first began serving as a statistical consultant almost forty years ago, I've made it a practice to begin every analysis by first computing the minimum and maximum of each variable. Can you tell why this practice would be of value to you as well?

12. Your mother has brought your attention to a newspaper article in which it is noted that one school has successfully predicted the outcome of every election of a U.S. president since 1976. Explain to her why this news does not surprise you.

13. A clinical study is well under way when it is noted that the values of critical end points vary far more from subject to subject than was expected originally. It is decided to increase the sample size. Is this an acceptable practice?

14. A clinical study is well under way when an unusual number of side effects are observed. The treatment code is broken and it is discovered that the majority of the effects are occurring in subjects in the control group. Two cases arise:

(a) The difference between the two treatment groups is statistically significant. It is decided to terminate the trials and recommend adoption of the new treatment. Is this an acceptable practice?

(b) The difference between the two treatment groups is *not* statistically significant. It is decided to continue the trials but to assign twice as many subjects to the new treatment as are placed in the control group. Is this an acceptable practice?

15. A jurist has asked for your assistance with a case involving possible racial discrimination. Apparently the passing rate of minorities was 90% compared to 97% for whites. The jurist didn't think this was much of a difference, but then one of the attorneys pointed out that these numbers represented a jump in the failure rate from 3% to 10%. How would you go about helping this jurist to reach a decision?

When you hired on as a statistician at the Bumbling Pharmaceutical Company, they told you they'd been waiting a long time to find a candidate like you. Apparently they had, for your desk is already piled high with studies that are long overdue for analysis. Here is just a sample:

16. The end-point values recorded by one physician are easily 10 times those recorded by all other investigators. Trying to track down the discrepancies, you discover that this physician has retired and closed his office. No one knows what became of his records. Your co-workers instantly begin to offer you advice including all of the following:

(a) Discard all the data from this physician.

(b) Assume this physician left out a decimal point and use the corrected values.

(c) Report the results for this observer separately.

(d) Crack the treatment code and then decide.

What will you do?

17. A different clinical study involved this same physician. This time, he completed the question about side effects that asked is this effect "mild, severe, or life threatening" but failed to answer the preceding question that specified the nature of the side effect. Which of the following should you do?

(a) Discard all the data from this physician.

(b) Discard all the side effect data from this physician.

(c) Report the results for this physician separately from the other results.

(d) Crack the treatment code and then decide.

18. Summarizing recent observations on the planetary systems of stars, the Monthly Notices of the Royal Astronomical Society reported that the vast majority of extrasolar planets in our galaxy must be gas giants like

Jupiter and Saturn as no Earth-size planet has been observed. What is your opinion?

9.2. SOLVING PRACTICAL PROBLEMS

In what follows, we suppose that you have been given a data set to analyze. The data did not come from a research effort that you designed, so there may be problems, many of them. We suggest you proceed as follows:

1. Determine the provenance of the observations.
2. Inspect the data.
3. Validate the data collection methods.
4. Formulate your hypotheses in testable form.
5. Choose methods for testing and estimation.
6. Be aware of what you don't know.
7. Perform the analysis.
8. Qualify your conclusions.

9.2.1. The Data's Provenance

Your very first questions should deal with *how* the data were collected. From what population(s) were they drawn? Were the members of the sample(s) selected at random? Were the observations independent of one another? If treatments were involved, were individuals assigned to these treatments at random? Remember, statistics is applicable only to random samples.[1] You need to find out all the details of the sampling procedure to be sure.

You also need to ascertain that the sample is representative of the population from which it purports to be drawn. If not, you'll need either to (1) weight the observations, (2) stratify the sample to make it more representative, or (3) redefine the population before drawing conclusions from the sample.

9.2.2. Inspect the Data

If satisfied with the data's provenance, you can now begin to inspect the data you've been provided. Your first step should be to compute the

---

[1] The one notable exception is that it is possible to make a comparison between entire populations by permutation means.

minimum and the maximum of each variable in the data set and to compare them with the data ranges you were provided by the client. If any lie outside the acceptable range, you need to determine which specific data items are responsible and have these inspected and, if possible, corrected by the person(s) responsible for their collection.

I once had a long-term client who would not let me look at the data. Instead, he would merely ask me what statistical procedure to use next. I ought to have complained, but this client paid particularly high fees, or at least he did so in theory. The deal was that I would get my money when the firm for which my client worked got its first financing from the venture capitalists. So my thoughts were on the money to come and not on the data.

My client took ill—later I was to learn he had checked into a rehabilitation clinic for a met-amphetamine addiction—and his firm asked me to take over. My first act was to ask for my money—they'd got their financing. While I waited for my check, I got to work, beginning my analysis as always by computing the minimum and the maximum of each variable. Many of the minimums were zero. I went to verify this finding with one of the technicians only to discover that zeros were well outside the acceptable range.

The next step was to look at the individual items in the database. There were zeros everywhere. In fact, it looked as if more than half the data were either zeros or repeats of previous entries. Before I could report these discrepancies to my client's boss, he called me in to discuss my fees.

"Ridiculous," he said. We did not part as friends. I almost regret not taking the time to tell him that half the data he was relying on did not exist. Tant pis. No, they are not still in business.

Not incidentally, the best cure for bad data is prevention. I strongly urge that all your data be entered directly into a computer so it can be checked and verified immediately upon entry. You don't want to be spending time tracking down corrections long after whoever entered 19.1 can remember whether the entry was supposed to be 1.91 or 9.1 or even 0.191.

9.2.3. Validate the Data Collection Methods

Few studies proceed exactly according to the protocol. Physicians switch treatments before the trial is completed. Sets of observations are missing or incomplete. A measuring instrument may have broken down midway through and been replaced by another, slightly different unit. Scoring methods were modified and observers provided with differing criteria employed. You need to determine the ways in which the protocol was modified and the extent and impact of such modifications.

A number of preventive measures may have been employed. For example, a survey may have included redundant questions as cross-checks. You need to determine the extent to which these preventive measures were successful. Was blinding effective? Or did observers crack the treatment code? You need to determine the extent of missing data and whether this was the same for all groups in the study. You may need to ask for additional data derived from follow-up studies of nonresponders and dropouts.

9.2.4. Formulate Hypotheses

All hypotheses must be formulated *before* the data are examined. It is all too easy for the human mind to discern patterns in what is actually a sequence of totally random events—think of the faces and animals that always seem to form in the clouds.

As another example, suppose that while just passing the time you deal out a five-card poker hand. It's a full house! Immediately, someone exclaims "What's the probability that could happen?" If by "that" a full house is meant, its probability is easily computed. But the same exclamation might have resulted had a flush or a straight been dealt or even three of a kind. The probability that "an interesting hand" will be dealt is much greater than the probability of a full house. Moreover, this might have been the third or even the fourth poker hand you've dealt; it's just that this one was the first to prove interesting enough to attract attention.

The details of translating objectives into testable hypotheses were given in Chapters 5 and 8.

9.2.5. Choosing a Statistical Methodology

For the two-sample comparison, a *t*-test should be used. Remember, one-sided hypotheses lead to one-sided tests and two-sided hypotheses to two-sided tests. If the observations were made in pairs, the paired *t*-test should be used.

Permutation methods should be used to make *k*-sample comparisons. Your choice of statistic will depend on the alternative hypothesis and the loss function.

Permutation methods should be used to analyze contingency tables.

The bootstrap is of value in obtaining confidence limits for quantiles and in model validation.

9.2.6. Be Aware of What You Don't Know

Far more statistical theory exists than can be provided in the confines of an introductory text. Entire books have been written on the topics of survey design, sequential analysis, and survival analysis and that's just the letter "s." If you are unsure what statistical method is appropriate, don't hesitate to look it up on the Web or in a more advanced text.

9.2.7. Qualify Your Conclusions

Your conclusions can only be applicable to the extent that samples were representative of populations and experiments and surveys were free from bias. A report by G. C. Bent and S. A. Archfield is ideal in this regard.[2] This report can be viewed online at `http://water.usgs.gov/pubs/wri/wri024043/`.

They devote multiple paragraphs to describing the methods used, the assumptions made, the limitations on their model's range of application, potential sources of bias, and the method of validation. For example: "The logistic regression equation developed is applicable for stream sites with drainage areas between 0.02 and 7.00 mi^2 in the South Coastal Basin and between 0.14 and 8.94 mi^2 in the remainder of Massachusetts, because these were the smallest and largest drainage areas used in equation development for their respective areas.

"The equation may not be reliable for losing reaches of streams, such as for streams that flow off area underlain by till or bedrock onto an area underlain by stratified-drift deposits. . . .

"The logistic regression equation may not be reliable in areas of Massachusetts where ground-water and surface-water drainage areas for a stream site differ." (Brent and Archfield provide examples of such areas.)

This report also illustrates how data quality, selection, and measurement bias can affect results. For example: "The accuracy of the logistic regression equation is a function of the quality of the data used in its development. This data includes the measured perennial or intermittent status of a stream site, the occurrence of unknown regulation above a site, and the measured basin characteristics.

"The measured perennial or intermittent status of stream sites in Massachusetts is based on information in the USGS NWIS database. Stream-flow measured as less than 0.005 ft^3/s is rounded down to zero, so it is possible that several streamflow measurements reported as zero may

---

[2] A logistic regression equation for estimating the probability of a stream flowing perennially in Massachusetts. USGC, Water-Resources Investigations Report 02-4043.

have had flows less than $0.005\,\text{ft}^3/\text{s}$ in the stream. This measurement would cause stream sites to be classified as intermittent when they actually are perennial."

It is essential that your reports be similarly detailed and qualified whether they are to a client or to the general public in the form of a journal article.

APPENDIX

S-PLUS

R and S-PLUS are both based on the same S language, and most programs written for one will work on the other.

This appendix describes two features unique to S-PLUS which may be of interest to readers of this book—the graphical user interface and the S+Resample library.

GRAPHICAL USER INTERFACE

Many of the options in S-PLUS are available from a graphical user interface (GUI). For example, you may inport data from a file or database using the `File : Import Data` menu, then browse to select the file to load.

Introduction to Statistics Through Resampling Methods and R/S-PLUS®, By Phillip I. Good
Copyright © 2005 by John Wiley & Sons, Inc.

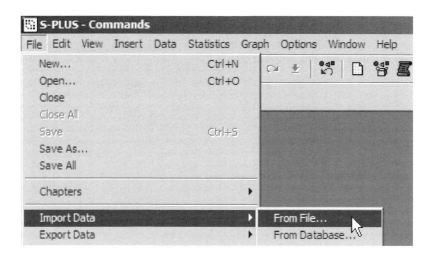

You may select a dataset to work with using the `Data : Load Data` menu; that brings up a spreadsheet window showing the data:

You may then select variables to plot and choose plots from the plot menu:

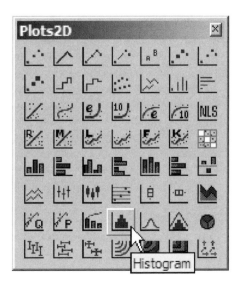

You may access statistical calculations from the menus. For example, the `File : Regression : Linear` menu brings up this menu, from which you may select the response and explanatory variables and set other options:

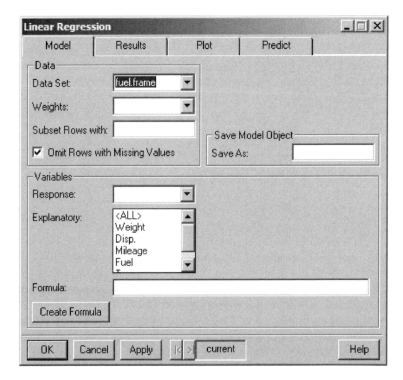

You can find other capabilities by exploring the menus, or in the online help:

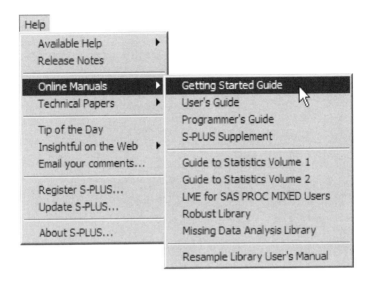

S+RESAMPLE LIBRARY

The S+Resample library may already be built into S-PLUS by the time you read this. If it is not, you may obtain it from `http://insightful.com/ downloads/libraries`.

There are installation instructions on that page. Installation is not difficult; if on Windows, you will download a resample.zip; you should double-click that to start whatever program your computer uses for zip files. This will create a resample folder; move this resample folder to the right place, typically something like C:/Program Files/Insightful/splus62/library/resample.

To use the library, use the `File : Load Library` menu, or load it from the command line using

```
library(resample, first = T)
```

The library makes it easy to do many of the most common resampling tasks. For example, the `Statistics : Resample : Two-sample t` menu brings up this menu, which we may use for bootstrap or permutation tests to compare the means of two groups; here the difference in means for variable *Time* between two groups in variable *group*, in data set *Verizon*.

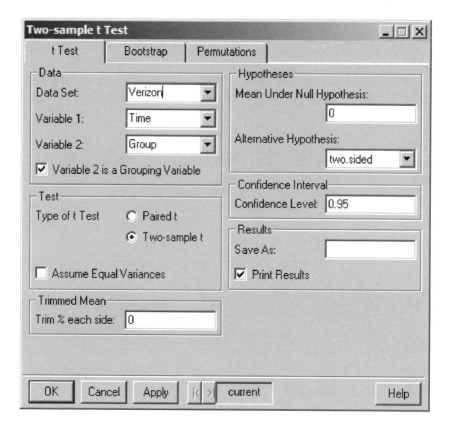

Bootstrap options are available from the bootstrap tab, shown here; similar options are shown for the permutation test.

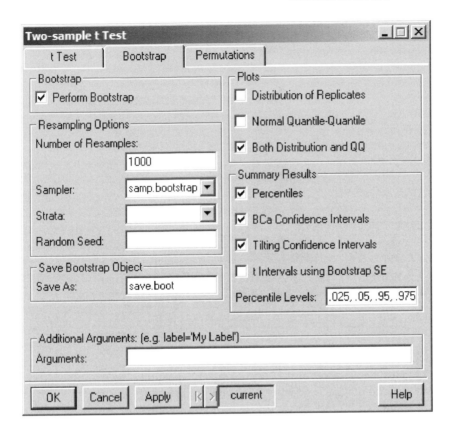

Here is one of the plots, showing the permutation distribution, with the observed value shown as a small vertical line segment on the left side of the distribution; the one-sided p-value is the area under the curve to the left of the observed value. This is the estimated fraction of the time that random sampling gives a result as or more extreme than the observed difference in means, assuming the null hypothesis is true.

permutation : Verizon$Time : mean : ILEC – CLEC

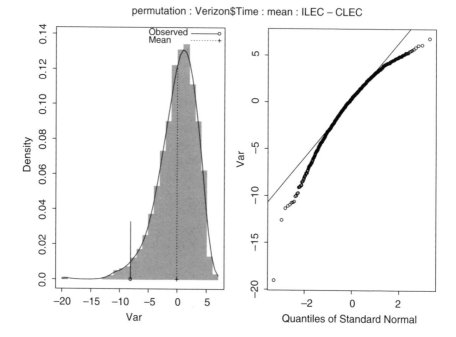

Finally, here is part of the printed output, including both results and the call to permutationTestMeans that you could use if working solely from the command line (here we set the random number seed to make results reproducible):

```
            *** Permutation Test Results ***
Call:
permutationTestMeans(data = Verizon$Time, treatment =
Verizon$Group, B = 999, seed = 0)
          Number of Replications: 999
          Summary Statistics:
     Observed      Mean      SE alternative p.value
Var    -8.098  -0.05759   3.235    two.sided   0.046
```

INDEX TO R COMMANDS

arithmetic functions
  abs(), 132
  prod(), 17
  round(), 41
  sqrt(), 167
  sum(), 103
assigning labels (name), 12
attach(), 24

binomial distribution (qbinom, dbinom, rbinom), 40, 115
boot.ci, 84
bootstrap sample, 20

controlling program flow
  for(), 20, 42, 182
  if, 37
  ifelse(), 105
  stop, 150

data.frame(), 198

exponential, 77

factor(), 13
file manipulations
  dir.create(), 131
  load(), 131
  read.table, 24
functions, creating, 37

graphics
  boxplot(), 8

coplot(), 199
histogram, 9
lines, 163
origin, 12
pie(), 12
plot(), 8
points(), 58
rug, 10
side by side, 199
stripchart(), 8

libraries
  boot, 83
  download and install, 83
  quantreg, 165
  stats, 85
  tree, 186

matrix, 146, 150
modeling functions
  lm()
    coef, 166
    fitted, 163
    summary, 167
    step, 175
  lsft(), 162
  rq(): coef, 182
  tree(), 187
    cut(), 188
    prune.tree(), 189

normal distribution (qnorm, dnorm, rnorm), 63, 89, 167

Introduction to Statistics Through Resampling Methods and R/S-PLUS®, By Phillip I. Good
Copyright © 2005 by John Wiley & Sons, Inc.

Poisson distribution (`qpois`, `dpois`, `rpois`), 59
programming guidelines, 149

random rearrangement, see `sample()`
reading data from a tabular file, 24
recursive function, 37

`sample()`, 20, 23, 72
statistics
  `cor()`, 93
  `max()`, 40
  `mean()`, 15
  `median()`, 5
  `quantile()`, 77
  `t.test()`, 79
subset selection, 198

trigonometric functions
  `cos()`, 159

user-created functions
  `comb()`, 141
  `eiv()`, 167
  `fact()`, 37
  `F1()`, 132
  how to, 37
  `pitcor.test()`, 136

vectors
  creating: numeric, seq
  `data.frame`, 198
  filling
    `factor()`, 13
    `numeric()`, 20
    `rep()`, 199
    `seq()`, 132
  matrix form, 25, 150
  `rank()`, 145
  `sort()`, 7

INDEX

Accuracy, 18, 77
Additive model, 158
Alternative hypothesis, 66, 73, 87, 107, 118, 133, 195
Analog vs. digital, 58
Association rule, 45
Assumptions, 89, 168
Audit, 22, 108, 166
Autoradiography, 66

Baseline, 138, 202
Bias, 206
Binomial
  distribution, 40, 63, 115
  parameter, 81
  probability, 38
  random variable, 58
  sample, 42
  trial, 31, 56, 80
Birth order, 28
Blinding, 97
Blocking, 104, 108, 194
Bootstrap, 88, 114, 186, 202, 215
  bias-corrected, 83, 85, 89
  parametric, 77, 86
  percentile, 18, 19, 71, 76
Box and whiskers plot, 6, 202
Box plot, 6, 8, 26
Boyle's Law, 1, 155

California Code of Civil Procedure (CCCP), 22
CART, *see* Classification and regression trees

Categorical variable, 13
Cauchy distribution, 65
Cause-and-effect, 156
Cell culture, 59, 66
Cell frequency, 142
Chi-square, 144, 148
Classification, 157
Classification and regression trees, 186
Coefficient, 155, 161, 167
Coin toss, 30
Combinations, 141
Complimentary probability, 34
Conditional probability, 43
Confidence, 45
Confidence interval, 71, 77, 81, 85, 95, 169, 203
Confound, 108
Contingency table, 11, 145
Controlling, 104
Controls, 98
Correlation, 72, 92, 112, 158
Criminal law, 51
Cross-validation, 186
Cumulative distribution, 10, 53, 58, 90, 92

Data
  categorical, 11, 137, 165
  continuous, 11, 55, 60
  discrete, 11, 55
  metric, 11, 165
  ordinal, 11
Data collection, 3, 13, 22, 194, 214
Debugging, 150

Introduction to Statistics Through Resampling Methods and R/S-PLUS®, By Phillip I. Good
Copyright © 2005 by John Wiley & Sons, Inc.

Decimal places, 201
Decisions, 16
Dice, 43
Distribution, 2
Dose response, 135

Effect size, 71
Election, 43, 70, 111
Empirical distribution, 42, 54, 120
Epidemic, 157
Equally likely events, 29
Estimation, 16, 18
Examples
    agriculture, 130, 137, 139, 172
    astrophysics, 58, 102
    biology, 91, 94, 99, 103, 105, 130, 134,
        151–152, 157, 162, 183, 191
    clinical trials, 88, 92, 98, 119, 126, 205
    economic, 45, 101, 108, 146, 151, 164,
        181
    geologic, 214
    law, 22, 38, 98, 102
Exchangeable observations, 89, 136
Expected value, 41, 61, 83
Experimental design, 2, 66, 98
Experimental unit, 106, 194
Exponential distribution, 61, 77, 148

Factorial, 35, 141
Fisher's exact test, 142
Fisher's omnibus statistic, 148

Geiger counter, 61
Genetics, 49
Goodness of fit, 170, 172
Graphs, 6, 8, 12, 26, 58, 164, 196, 199,
    202
Group sequential design, 122
Growth processes, 129, 200
Guidelines, 149, 170

Histogram, 9, 15, 198
HIV, 107
Homogeneity, 14, 73
Hypothesis
    formulation, 100, 106, 193, 215
    testing, 66, 204

Identically distributed, 89
Independence, 47–49, 64, 100, 111

Independent events, 39, 111
Indicator function, 148
Intercept, 169
Interquartile range, 6, 21, 27, 77

Jury selection, 22, 38, 109

Key, 186

Likert scale, 154, 165, 173
Logarithms, 130, 164
Losses, 70, 114, 57
LSAT, 28, 94

Marginals, 140–142
Martingale, 30
Matched pairs, 104
Matrix, 147
Maximum, 5, 10
Mean
    arithmetic, 14, 16, 90
    geometric, 17, 200
Measurement errors, 2
Median, 4, 8, 14–17, 181, 200
Meta-analysis, 126
Minimum, 5, 10
Missing data, 202, 206
Mode, 15, 16, 205
Model, 155
Model order, 157
Moiser's method, 185
Monotone function, 61, 157
Monte Carlo, 81, 90, 92
Multinomial, 144
Multiple tests, 71
Multisample comparison, 131, 136
Mutually exclusive events, 31

Nonrespondents, 196, 205
Nonsymmetrical, *see* Skewed
Normal distribution, 25, 62, 114, 117,
    198
Null hypothesis, 67, 80, 140, 143, 148

Objectives, 193
Observer error, 3
One- vs. two-sided, 74, 81, 83, 87, 106,
    143, 195
Ordered samples, 134, 136
Outcomes vs. events, 31

Parameter, 16, 62, 114, 160, 197
Patterns, 8
Pearson correlation, 93
Percentages, 129
Percentile, 6, 9, 77, 90, 181, 216
Permutation, 35, 93
Permutation test, 82, 88, 95, 131, 134, 215
Petri dish, 58
Pie chart, 11
Pitman correlation, 92
Placebo effect, 97
Poisson distribution, 58, 80, 201
Population, 19, 101, 103, 197
Power, 91, 114, 117, 119, 193, 195
Precision, 18, 197
Prediction, 156, 170
Predictor, 155, 161, 173
Probability laws, 31
Prune, 188
p-value, 82, 88, 93, 126, 146, 203

Quantile, *see* Percentile
Quetlet index, 165

Random numbers, 22
Randomize, 104
Range, 5, 26
Ranks, 138–139, 148
Read files, 24
Rearrangement, 36, 68, 83, 94, 105, 138
Recursion, 141
Regression, 158
    Deming, 166
    least absolute deviation (LAD), 165
    linear, 160
    multivariable, 172
    nonlinear, 161
    ordinary least-squares (OLS), 161
    quantile, 179
    stepwise, 172–177
Regression tree, *see* Classification and regression trees (CART)
Reportable elements, 192
Resample, 149
Residual, 162, 169, 171
Response variable, 155, 161
Robust, 17, 90
Roulette wheel, 30
R-squared, 175

Sample
    random, 22, 52, 66, 96, 100, 110
    representative, 21, 25, 108, 113, 216
    size, 18, 110–125, 193
    splitting, 184, 190
Sampling, 19, 21, 82, 194
    adaptive, 126
    clusters, 111, 195
    sequential, 113, 120
    unit, 23
Scatter plot, 12, 164, 199
Selectivity, 92
Sensitivity, 92
Shift alternative, 54, 67
Significance level, 75, 88, 91, 114, 120, 203
Significant digits, 9
Simulation, 82, 114, 120
Skewed, 17, 201
Slope, 167
Sort, 7
Spreadsheet, 23
Standard deviation, 25, 78, 118
Standard error, 86, 201
Statistics, 15, 62, 131
Stein's two-stage procedure, 120
Strata, *see* Blocking
Strip chart, 5, 8, 10, 15, 20
Student's *t*, 78, 86, 90, 92, 148, 215
Sufficient statistic, 58
Summarize, 4
Support, 45
Surrogate, 107
Survey, 4, 21, 99, 108
Survival probability, 122, 140
Symmetric, 57, 62

Terminal node, 188
Treatment allocation, 109
Type I, II errors, 69, 79, 88, 91, 114–116

Uniform distribution, 201

Vaccine, 80
Validation, 158, 183, 195
Variance, 57, 65, 87, 90, 117, 137
Variation, 1, 25, 41, 100, 103, 114, 194
Venn diagram, 31, 44, 48
Vitamin E, 67